INTERNATIONAL CODES™

ISPSC®

A Member of the International Code Family®

INCLUDES
APSP-7 Standard for Suction Entrapment Avoidance

INTERNATIONAL **SWIMMING POOL AND SPA** CODE®

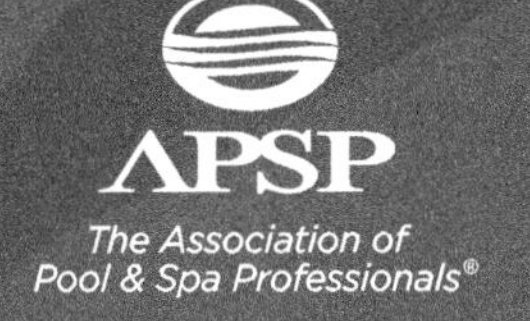

2018 International Swimming Pool and Spa Code®

First Printing: August 2017
Second Printing: May 2018

ISBN: 978-1-60983-746-4 (soft-cover edition)

Date of First Publication: August 31, 2017

T022589

PRINTED IN THE U.S.A.

PREFACE

Introduction

The *International Swimming Pool and Spa Code*® (ISPSC®) establishes minimum requirements for the design, construction, alteration, repair and maintenance of swimming pools, spas, hot tubs and aquatic facilities. This 2018 edition is fully compatible with all of the *International Codes*® (I-Codes®) published by the International Code Council® (ICC®), including the *International Building Code*®, *International Energy Conservation Code*®, *International Existing Building Code*®, *International Fire Code*®, *International Fuel Gas Code*®, *International Green Construction Code*®, *International Mechanical Code*®, *International Plumbing Code*®, *International Private Sewage Disposal Code*®, *International Property Maintenance Code*®, *International Residential Code*®, *International Wildland-Urban Interface Code*®, *International Zoning Code*® and *International Code Council Performance Code*®.

The I-Codes, including this *International Swimming Pool and Spa Code*, are used in a variety of ways in both the public and private sectors. Most industry professionals are familiar with the I-Codes as the basis of laws and regulations in communities across the U.S. and in other countries. However, the impact of the codes extends well beyond the regulatory arena, as they are used in a variety of nonregulatory settings, including:

- Voluntary compliance programs such as those promoting sustainability, energy efficiency and disaster resistance.
- The insurance industry, to estimate and manage risk, and as a tool in underwriting and rate decisions.
- Certification and credentialing of individuals involved in the fields of building design, construction and safety.
- Certification of building and construction-related products.
- U.S. federal agencies, to guide construction in an array of government-owned properties.
- Facilities management.
- "Best practices" benchmarks for designers and builders, including those who are engaged in projects in jurisdictions that do not have a formal regulatory system or a governmental enforcement mechanism.
- College, university and professional school textbooks and curricula.
- Reference works related to building design and construction.

In addition to the codes themselves, the code development process brings together building professionals on a regular basis. It provides an international forum for discussion and deliberation about building design, construction methods, safety, performance requirements, technological advances and innovative products.

Development

This 2018 edition presents the code as originally issued, with changes as reflected in the 2015 edition and further changes approved by the ICC Code Development Process through 2017. A new edition such as this is promulgated every 3 years.

This code is founded on principles intended to establish provisions consistent with the scope of a swimming pool and spa code that adequately protects public health, safety and welfare; provisions that do not unnecessarily increase construction costs; provisions that do not restrict the use of new materials, products or methods of construction; and provisions that do not give preferential treatment to particular types or classes of materials, products or methods of construction.

Maintenance

The *International Swimming Pool and Spa Code* is kept up to date through the review of proposed changes submitted by code enforcement officials, industry representatives, design professionals and other interested parties. Proposed changes are carefully considered through an open code development process in which all interested and affected parties may participate.

The ICC Code Development Process reflects principles of openness, transparency, balance, due process and consensus, the principles embodied in OMB Circular A-119, which governs the federal government's use of private-sector standards. The ICC process is open to anyone; there is no cost to participate, and people can participate without travel cost through the ICC's cloud-based app, cdpAccess®. A broad cross section of interests are represented in the ICC Code Development Process. The codes, which are updated regularly, include safeguards that allow for emergency action when required for health and safety reasons.

In order to ensure that organizations with a direct and material interest in the codes have a voice in the process, the ICC has developed partnerships with key industry segments that support the ICC's important public safety mission. Some code development committee members were nominated by the following industry partners and approved by the ICC Board:

- Association of Pool and Spa Professionals (APSP)
- American Institute of Architects (AIA)

The code development committees evaluate and make recommendations regarding proposed changes to the codes. Their recommendations are then subject to public comment and council-wide votes. The ICC's governmental members—public safety officials who have no financial or business interest in the outcome—cast the final votes on proposed changes.

The contents of this work are subject to change through the code development cycles and by any governmental entity that enacts the code into law. For more information regarding the code development process, contact the Codes and Standards Development Department of the International Code Council.

While the I-Code development procedure is thorough and comprehensive, the ICC, its members and those participating in the development of this code disclaim any liability resulting from the publication or use of the I-Codes, or from compliance or noncompliance with their provisions. The ICC does not have the power or authority to police or enforce compliance with the contents of this code.

Code Development Committee Responsibilities (Letter Designations in Front of Section Numbers)

In each code development cycle, code change proposals to this code are considered at the Committee Action Hearing by the International Swimming Pool and Spa Code Development Committee, whose action constitutes a recommendation to the voting membership for final action on the proposed change. Code change proposals to sections of the code that are preceded by a bracketed letter designation, such as [A], will be considered by a committee other than the Swimming Pool and Spa Code Development Committee. For example, proposed changes to Section [A] 102.1 will be considered by the Administrative Code Development Committee during the Committee Action Hearings in the 2019 (Group B) code development cycle.

The bracketed letter designations for committees responsible for portions of this code are as follows:

[A] = Administrative Code Development Committee; and

[BS] = IBC—Structural Code Development Committee

For the development of the 2021 edition of the I-Codes, there will be two groups of code development committees and they will meet in separate years. Note that these are tentative groupings.

Group A Codes (Heard in 2018, Code Change Proposals Deadline: January 8, 2018)	Group B Codes (Heard in 2019, Code Change Proposals Deadline: January 7, 2019)
International Building Code – Egress (Chapters 10, 11, Appendix E) – Fire Safety (Chapters 7, 8, 9, 14, 26) – General (Chapters 2–6, 12, 27–33, Appendices A, B, C, D, K, N)	Administrative Provisions (Chapter 1 of all codes except IECC, IRC and IgCC, administrative updates to currently referenced standards, and designated definitions)
International Fire Code	**International Building Code** – Structural (Chapters 15–25, Appendices F, G, H, I, J, L, M)
International Fuel Gas Code	**International Existing Building Code**
International Mechanical Code	**International Energy Conservation Code—Commercial**
International Plumbing Code	**International Energy Conservation Code—Residential** – IECC—Residential – IRC—Energy (Chapter 11)
International Property Maintenance Code	**International Green Construction Code** (Chapter 1)
International Private Sewage Disposal Code	**International Residential Code** – IRC—Building (Chapters 1–10, Appendices E, F, H, J, K, L, M, O, Q, R, S, T)
International Residential Code – IRC—Mechanical (Chapters 12–23) – IRC—Plumbing (Chapters 25–33, Appendices G, I, N, P)	
International Swimming Pool and Spa Code	
International Wildland-Urban Interface Code	
International Zoning Code	
Note: Proposed changes to the ICC *Performance Code*™ will be heard by the code development committee noted in brackets [ ] in the text of the ICC *Performance Code*™.	

Code change proposals submitted for code sections that have a letter designation in front of them will be heard by the respective committee responsible for such code sections. Because different committees hold Committee Action Hearings in different years, proposals for this code will be heard by committees in both the 2018 (Group A) and the 2019 (Group B) code development cycles.

Note that every section of Chapter 1 of this code is designated as the responsibility of the Administrative Code Development Committee, and that committee is part of the Group B portion of the code hearings. This committee will hold its code development hearings in 2019 to consider all code change proposals for Chapter 1 of this code and proposals for Chapter 1 of all I-Codes except the *International Energy Conservation Code*, *International Residential Code* and *International Green Construction Code*. Therefore, any proposals received for Chapter 1 of this code will be assigned to the Administrative Code Development Committee for consideration in 2019.

It is very important that anyone submitting code change proposals understand which code development committee is responsible for the section of the code that is the subject of the code change proposal. For further information on the code development committee responsibilities, please visit the ICC website at www.iccsafe.org/scoping.

Marginal Markings

Solid vertical lines in the margins within the body of the code indicate a technical change from the requirements of the 2015 edition. Deletion indicators in the form of an arrow (➡) are provided in the margin where an entire section, paragraph, exception or table has been deleted or an item in a list of items or a table has been deleted.

Coordination of the International Codes

The coordination of technical provisions is one of the strengths of the ICC family of model codes. The codes can be used as a complete set of complementary documents, which will provide users with full integration and coordination of technical provisions. Individual codes can also be used in subsets or as stand-alone documents. To make sure that each individual code is as complete as possible, some technical provisions that are relevant to more than one subject area are duplicated in some of the model codes. This allows users maximum flexibility in their application of the I-Codes.

Italicized Terms

Words and terms defined in Chapter 2, Definitions, are italicized where they appear in code text and the Chapter 2 definition applies. Where such words and terms are not italicized, common-use definitions apply. The words and terms selected have code-specific definitions that the user should read carefully to facilitate better understanding of the code.

Adoption

The International Code Council maintains a copyright in all of its codes and standards. Maintaining copyright allows ICC to fund its mission through sales of books, in both print and electronic formats. The ICC welcomes adoption of its codes by jurisdictions that recognize and acknowledge the ICC's copyright in the code, and further acknowledge the substantial shared value of the public/private partnership for code development between jurisdictions and the ICC.

The ICC also recognizes the need for jurisdictions to make laws available to the public. All I-Codes and I-Standards, along with the laws of many jurisdictions, are available for free in a nondownloadable form on the ICC's website. Jurisdictions should contact the ICC at adoptions@iccsafe.org to learn how to adopt and distribute laws based on the *International Swimming Pool and Spa Code* in a manner that provides necessary access, while maintaining the ICC's copyright.

To facilitate adoption, several sections of this code contain blanks for fill-in information that needs to be supplied by the adopting jurisdiction as part of the adoption legislation. For this code, please see:

Section 101.1. Insert: [NAME OF JURISDICTION]

Section 105.6.2. Insert: [APPROPRIATE SCHEDULE]

Section 105.6.3: [PERCENTAGE IN TWO LOCATIONS]

Section 107.4. Insert: [OFFENSE]

Section 107.4. Insert: [DOLLAR AMOUNT]

Section 107.4. Insert: [NUMBER OF DAYS]

Section 107.5. Insert: [DOLLAR AMOUNT IN TWO LOCATIONS]

EFFECTIVE USE OF THE INTERNATIONAL SWIMMING POOL AND SPA CODE

The *International Swimming Pool and Spa Code* (ISPSC) is a model code that regulates the minimum requirements for the design, construction, alteration, repair and maintenance of swimming pools, spas, hot tubes and aquatic facilities. This includes public swimming pools, public spas, public exercise spas, aquatic recreation facilities, onground storable residential pools, permanent inground residential pools, permanent residential spas, permanent residential exercise spas, portable residential spas and portable residential exercise spas.

In many jurisdictions, in addition to code officials having the responsibility for reviewing plans and inspecting the construction of pools and spas, environmental health officials also have a responsibility for oversight of the operation of pools and spas. In order to prevent disease and prevent injuries, environmental health officials conduct operational evaluations (inspections). This may include water chemistry, credentials and training of pool operators and lifeguards, proper water circulation, facility staff's preparedness to respond to injuries and accidents, and proper sanitation and safety of the facility.

Code officials and environmental health officials commonly work closely in the plan review and inspection of pools and spas. This collaboration between departments to jointly review plans and inspect pools and spas is critical in order to achieve a safe and healthy environment for all that utilize these facilities.

The Association of Pool & Spa Professionals (APSP), a cooperating sponsor with ICC in the development and update of the ISPSC, further notes: "While it is recognized that proper construction and installation are essential, safe use of pools and spas requires common sense, including constant adult supervision of children, and proper maintenance. It is assumed and intended that pool users will exercise appropriate personal judgment and responsibility (including constant adult supervision of children) and that operators will create and enforce rules and warning appropriate for their pool/spa."

Arrangement and Format of the 2018 ISPSC

The format of the ISPSC allows each chapter to be devoted to a particular subject with the exception of Chapter 3 which contains general compliance subject matter that is coordinated with the provisions for each type of pool and spa regulated in Chapters 4–10. The ISPSC is divided into eleven different parts:

Chapter	Subject
1	Scope and Administration
2	Definitions
3	General Compliance
4	Public Swimming Pools
5	Public Spas and Public Exercise Spas
6	Aquatic Recreation Facilities
7	Onground Storable Residential Swimming Pools
8	Permanent Inground Residential Swimming Pools
9	Permanent Residential Spas and Permanent Residential Exercise Spas
10	Portable Residential Spas and Portable Residential Exercise Spas
11	Referenced Standards

The following is a chapter-by-chapter synopsis of the scope and intent of the provisions of the *International Swimming Pool and Spa Code*:

Chapter 1 Scope and Administration. This chapter contains provisions for the application, enforcement and administration of subsequent requirements of the code. Chapter 1 identifies which swimming pools and spas come under its purview. It is largely concerned with maintaining "due process of law" in enforcing the design and construction criteria contained in the body of the code. Only through careful observation of the administrative provisions can the code official reasonably expect to demonstrate that "equal protection under the law" has been provided.

Chapter 2 Definitions. Terms that are defined in the code are listed alphabetically in Chapter 2. While a defined term may be used in one chapter or another, the meaning provided in Chapter 2 is applicable throughout the code.

Where understanding of a term's definition is especially key to or necessary for the understanding of a particular code provision, the term is shown in *italics* wherever it appears in the code. This is true only for those terms that have a meaning that is unique to the code. In other words, the generally understood meaning of a term or phrase might not be sufficient or consistent with the meaning prescribed by the code; therefore, it is essential that the code-defined meaning be known.

Guidance regarding tense, gender and plurality of defined terms as well as guidance regarding terms not defined in this code is provided.

Chapter 3 General Compliance. Chapter 3 is broad in scope. It includes a variety of requirements for pools and spas. This chapter provides requirements that are intended to maintain a minimum level of safety and sanitation for both the general public and the users of pools or spas. Chapter 3 provides specific criteria for electrical, plumbing, mechanical and fuel gas requirements; energy savings requirements; construction in flood hazard areas; barrier requirements; decks around pools and spas; general design; dimensional design; equipment; suction entrapment avoidance; circulation systems; filters; pumps and motors; return and suction fittings; skimmers; heaters; air blowers and air induction systems; water supply; sanitizing equipment; wastewater disposal; lighting; ladders and recessed treads; and safety. It is important to note that Chapter 3 is intended to provide general requirements not found in Chapters 4–10. Chapters 4–10 specifically reference Chapter 3 in order to coordinate the general provisions with the specific provisions based on the type of pool or spa.

Chapter 4 Public Swimming Pools. The purpose of Chapter 4 is to set forth specific requirements in the code for public swimming pools with regard to diving equipment, bather load limitations, rest ledges, wading pools, decks, deck equipment, filters, dressing and sanitary facilities, special features and signage. The term "public swimming pool" is defined in Chapter 2 and includes the different classes of pools (Class A – Class F).

Chapter 5 Public Spas and Public Exercise Spas. Chapter 5 establishes the specific criteria for public spas and public exercise spas with regard to materials, structure and design, pumps and motors, return and suction fittings, heater and temperature requirements, water supply, sanitation, oxidation equipment and chemical feeders, and safety features. The term "spa" is defined in Chapter 2.

Chapter 6 Aquatic Recreation Facilities. The purpose of Chapter 6 is to establish specific requirements for aquatic recreation facilities with regard to floors, markings and indications, circulation systems, handholds and ropes, depths, barriers, number of occupants, toilet rooms and bathrooms, special features and signage. The term "aquatic recreation facility" is defined in Chapter 2 and includes wave pools, leisure rivers, inner tube rides and body slides, to name a few.

Chapter 7 Onground Storable Residential Swimming Pools. The purpose of Chapter 7 is to establish specific requirements for onground storable residential swimming pools with regard to ladders and stairs, decks and circulation systems. The term "onground storable pool" is defined in Chapter 2. This chapter applies to what has been commonly referred to in past standards and codes as onground or above-ground pools. The application of the provisions for onground residential pools is limited to pools associated with detached one- and two-family dwellings and townhouses not more than three stories high in accordance with the definition of the term "residential" in Chapter 2.

Chapter 8 Permanent Inground Residential Swimming Pools. The purpose of Chapter 8 is to establish specific requirements for permanent inground residential swimming pools with regard to design, construction tolerances, diving water envelopes, walls, offset ledges, pool floors, diving equipment, special features, circulation systems and safety features. The application of the provisions for inground residential pools is limited to pools associated with detached one- and two-family dwellings and townhouses not more than three stories high in accordance with the definition of the term "residential" in Chapter 2.

Chapter 9 Permanent Residential Spas and Permanent Residential Exercise Spas. The purpose of Chapter 9 is to establish specific requirements for permanent residential spas and permanent residential exercise spas with regard to safety features. The application of the provisions for residential spas ("spa" is defined in Chapter 2) is limited to spas associated with detached one- and two-family dwellings and townhouses not more than three stories high in accordance with the definition of the term "residential" in Chapter 2.

Chapter 10 Portable Residential Spas and Portable Residential Exercise Spas. The purpose of Chapter 10 is to establish specific requirements for portable residential spas and portable residential exercise spas with regard to standards that the equipment must meet. The application of the provisions for residential spas ("spa" is defined in Chapter 2) is limited to spas associated with detached one- and two-family dwellings and townhouses not more than three stories high in accordance with the definition of the term "residential" in Chapter 2.

Chapter 11 Referenced Standards. The code contains numerous references to standards that are used to regulate materials and methods of construction. Chapter 11 contains a comprehensive list of all standards that are referenced in the code. The standards are part of the code to the extent of the reference to the standard. Compliance with the referenced standard is necessary for compliance with this code. By providing specifically adopted standards, the construction and installation requirements necessary for compliance with the code can be readily determined. The basis for code compliance is, therefore, established and available on an equal basis to the code official, contractor, designer and owner.

Chapter 11 is organized in a manner that makes it easy to locate specific standards. It lists all of the referenced standards, alphabetically, by acronym of the promulgating agency of the standard. Each agency's standards are then listed in either alphabetical or numeric order based on the standard identification. The list also contains the title of the standard; the edition (date) of the standard referenced; any addenda included as part of the ICC adoption; and the section or sections of this code that reference the standard.

TABLE OF CONTENTS

CHAPTER 1

SCOPE AND ADMINISTRATION

User note:

About this chapter: *Chapter 1 establishes the limits of applicability of this code and describes how the code is to be applied and enforced. Chapter 1 is in two parts: Part 1—Scope and Application (Sections 101–102) and Part 2—Administration and Enforcement (Sections 103–108). Section 101 identifies which buildings and structures come under its purview and references other I-Codes an applicable. Standards and codes are scoped to the extent referenced (see Section 102.8).*

This code is intended to be adopted as a legally enforceable document and it cannot be effective without adequate provisions for its administration and enforcement. The provisions of Chapter 1 establish the authority and duties of the code official appointed by the authority having jurisdiction and also establish the rights and privileges of the design professional, contractor and property owner.

PART 1—SCOPE AND APPLICATION

SECTION 101
GENERAL

[A] 101.1 Title. These regulations shall be known as the Swimming Pool and Spa Code of **[NAME OF JURISDICTION]**, hereinafter referred to as "this code."

[A] 101.2 Scope. The provisions of this code shall apply to the construction, alteration, movement, renovation, replacement, repair and maintenance of aquatic recreation facilities, pools and spas. The pools and spas covered by this code are either permanent or temporary, and shall be only those that are designed and manufactured to be connected to a circulation system and that are intended for swimming, bathing or wading.

101.2.1 Flotation tanks. Flotation tank systems intended for sensory deprivation therapy shall not be considered to be included in the scope of this code.

[A] 101.3 Intent. The purpose of this code is to establish minimum standards to provide a reasonable level of safety and protection of health, property and public welfare by regulating and controlling the design, construction, installation, quality of materials, location and maintenance or use of pools and spas.

[A] 101.4 Severability. If any section, subsection, sentence, clause or phrase of this code is for any reason held to be unconstitutional, such decision shall not affect the validity of the remaining portions of this code.

SECTION 102
APPLICABILITY

[A] 102.1 General. Where there is a conflict between a general requirement and a specific requirement, the specific requirement shall govern. Where, in any specific case, different sections of this code specify different materials, methods of construction or other requirements, the most restrictive shall govern.

[A] 102.2 Existing installations. Any pool or spa and related mechanical, electrical and plumbing systems lawfully in existence at the time of the adoption of this code shall be permitted to have their use and maintenance continued if the use, maintenance or repair is in accordance with the original design and no hazard to life, health or property is created.

[A] 102.3 Maintenance. Pools and spas and related mechanical, electrical and plumbing systems, both existing and new, and parts thereof, shall be maintained in proper operating condition in accordance with the original design in a safe and sanitary condition. Devices or safeguards that are required by this code shall be maintained in compliance with the edition of the code under which they were installed.

The owner or the owner's authorized agent shall be responsible for maintenance of systems. To determine compliance with this provision, the code official shall have the authority to require any system to be reinspected.

[A] 102.4 Additions, alterations or repairs. Additions, *alterations*, renovations or *repairs* to any pool, spa or related system shall conform to that required for a new system without requiring the existing systems to comply with the requirements of this code. Additions, alterations or repairs shall not cause existing systems to become unsafe, insanitary or overloaded.

Minor additions, alterations, renovations and repairs to existing systems shall be permitted in the same manner and arrangement as in the existing system, provided that such repairs or replacement are not hazardous and are *approved*.

[A] 102.5 Historic buildings. The provisions of this code relating to the construction, alteration, repair, enlargement, restoration, relocation or moving of pools, spas or systems shall not be mandatory for existing pools, spas or systems identified and classified by the state or local jurisdiction as part of a historic structure where such pools, spas or systems are judged by the code official to be safe and in the public interest of health, safety and welfare regarding any proposed construction, alteration, repair, enlargement, restoration, relocation or moving of such pool or spa.

[A] 102.6 Moved pools and spas. Except as determined by Section 102.2, systems that are a part of a pool, spa or system moved into or within the jurisdiction shall comply with the provisions of this code for new installations.

[A] 102.7 Referenced codes and standards. The codes and standards referenced in this code shall be those that are listed in Chapter 11 and such codes and standards shall be considered to be part of the requirements of this code to the prescribed extent of each such reference. Where differences occur between provisions of this code and the referenced standards, the provisions of this code shall be the minimum requirements.

[A] 102.7.1 Application of the International Codes. Where the *International Residential Code* is referenced in this code, the provisions of the *International Residential Code* shall apply to related systems in detached one- and two-family dwellings and townhouses not more than three stories in height. Other related systems shall comply with the applicable International Code or referenced standard.

[A] 102.8 Requirements not covered by code. Any requirements necessary for the strength, stability or proper operation of an existing or proposed system, or for the public safety, health and general welfare, not specifically covered by this code shall be determined by the code official.

[A] 102.9 Other laws. The provisions of this code shall not be deemed to nullify any provisions of local, state or federal law.

[A] 102.10 Application of references. References to chapter or section numbers, or to provisions not specifically identified by number, shall be construed to refer to such chapter, section or provision of this code.

PART 2—ADMINISTRATION AND ENFORCEMENT

SECTION 103 DEPARTMENT OF BUILDING SAFETY

[A] 103.1 Creation of enforcement agency. The department of building safety is hereby created and the official in charge thereof shall be known as the code official.

[A] 103.2 Appointment. The code official shall be appointed by the chief appointing authority of the jurisdiction.

[A] 103.3 Deputies. In accordance with the prescribed procedures of the jurisdiction and with the concurrence of the appointing authority, the code official shall have the authority to appoint a deputy code official, the related technical officers, inspectors, plans examiners and other employees. Such employees shall have powers as delegated by the code official.

[A] 103.4 Liability. The code official, member of the board of appeals or employee charged with the enforcement of this code, while acting for the jurisdiction in good faith and without malice in the discharge of the duties required by this code or other pertinent law or ordinance, shall not thereby be rendered civilly or criminally liable personally and is hereby relieved from personal liability for any damage accruing to persons or property as a result of any act or by reason of an act or omission in the discharge of official duties.

[A] 103.4.1 Legal defenses. Any suit or criminal complaint instituted against an officer or employee because of an act performed by that officer or employee in the lawful discharge of duties and under the provisions of this code shall be defended by legal representatives of the jurisdiction until the final termination of the proceedings. The code official or any subordinate shall not be liable for cost in any action, suit or proceeding that is instituted in pursuance of the provisions of this code.

SECTION 104 DUTIES AND POWERS OF THE CODE OFFICIAL

[A] 104.1 General. The code official is hereby authorized and directed to enforce the provisions of this code. The code official shall have the authority to render interpretations of this code and to adopt policies and procedures in order to clarify the application of its provisions. Such interpretations, policies and procedures shall be in compliance with the intent and purpose of this code. Such policies and procedures shall not have the effect of waiving requirements specifically provided for in this code.

[A] 104.2 Applications and permits. The code official shall receive applications, review construction documents and issue permits for the erection, alteration, demolition and moving of pools, spas and related mechanical, electrical and plumbing systems. The code official shall inspect the premises for which such permits have been issued and enforce compliance with the provisions of this code.

[A] 104.3 Notices and orders. The code official shall issue necessary notices or orders to ensure compliance with this code.

[A] 104.4 Inspections. The code official shall make the required inspections, or the code official shall have the authority to accept reports of inspection by *approved* agencies or individuals. Reports of such inspections shall be in writing and be certified by a responsible officer of such *approved* agency or by the responsible individual. The code official is authorized to engage such expert opinion as deemed necessary to report on unusual technical issues that arise, subject to the approval of the appointing authority.

[A] 104.5 Identification. The code official shall carry proper identification when inspecting structures or premises in the performance of duties under this code.

[A] 104.6 Right of entry. Where it is necessary to make an inspection to enforce the provisions of this code, or where the code official has reasonable cause to believe that there exists in a structure or on a premises a condition that is contrary to or in violation of this code that makes the structure or premises unsafe, dangerous or hazardous, the code official is authorized to enter the structure or premises at reasonable times to inspect or to perform the duties imposed by this code, provided that if such structure or premises be occupied that credentials be presented to the occupant and entry requested. If such structure or premises is unoccupied, the code official shall first make a reasonable effort to locate the owner, the owner's authorized agent or other person having charge or control of the structure or premises and request entry. If entry is refused, the code official shall have recourse to the remedies provided by law to secure entry.

[A] 104.7 Department records. The code official shall keep official records of applications received, permits and certificates issued, fees collected, reports of inspections, and notices and orders issued. Such records shall be retained in the official records for the period required for retention of public records.

[A] 104.8 Modifications. Where there are practical difficulties involved in carrying out the provisions of this code, the code official shall have the authority to grant modifications for individual cases, upon application of the owner or owner's authorized agent, provided that the code official shall first find that special individual reason makes the strict letter of this code impractical and the modification is in compliance with the intent and purpose of this code and that such modification does not lessen sustainability, health, accessibility, life safety and structural requirements. The details of action granting modifications shall be recorded and entered in the files of the department of building safety.

[A] 104.9 Alternative materials, design and methods of construction and equipment. The provisions of this code are not intended to prevent the installation of any design or material or to prohibit any method of construction not specifically prescribed by this code, provided that any such alternative has been *approved*. An alternative material, design or method of construction shall be *approved* where the code official finds that the proposed design is satisfactory and complies with the intent of the provisions of this code, and that the material, method or work offered is, for the purpose intended, not less than the equivalent of that prescribed in this code in quality, strength, effectiveness, durability and safety. Where the alternative material, design or method of construction is not *approved*, the *code official* shall respond in writing, stating the reasons why the alternative was not *approved*.

[A] 104.10 Required testing. Where there is insufficient evidence of compliance with the provisions of this code, or evidence that a material or method does not conform to the requirements of this code, or in order to substantiate claims for alternative materials or methods, the code official shall have the authority to require tests as evidence of compliance to be made at no expense to the jurisdiction.

[A] 104.10.1 Test methods. Test methods shall be as specified in this code or by other recognized test standards. In the absence of recognized and accepted test methods, the code official shall approve the testing procedures.

[A] 104.10.2 Testing agency. Tests shall be performed by an *approved* agency.

[A] 104.10.3 Test reports. Reports of tests shall be retained by the code official for the period required for retention of public records.

[A] 104.11 Alternative engineered design. The design, documentation, inspection, testing and approval of an alternative engineered design shall comply with Sections 104.11.1 through 104.11.6.

[A] 104.11.1 Design criteria. An alternative engineered design shall conform to the intent of the provisions of this code and shall provide an equivalent level of quality, strength, effectiveness, durability and safety. Material, equipment or components shall be designed and installed in accordance with the manufacturer's instructions.

[A] 104.11.2 Submittal. The registered design professional shall indicate on the permit application that the system is an alternative engineered design. The permit and permanent permit records shall indicate that an alternative engineered design was part of the *approved* installation.

[A] 104.11.3 Technical data. The registered design professional shall submit sufficient technical data to substantiate the proposed alternative engineered design and to prove that the performance meets the intent of this code.

[A] 104.11.4 Construction documents. The registered design professional shall submit to the code official two complete sets of signed and sealed construction documents for the alternative engineered design.

[A] 104.11.5 Design approval. Where the code official determines that the alternative engineered design conforms to the intent of this code, the system shall be *approved*. If the alternative engineered design is not *approved*, the code official shall notify the registered design professional in writing, stating the reasons why the alternative was not *approved*.

[A] 104.11.6 Inspection and testing. The alternative engineered design shall be tested and inspected in accordance with the requirements of Section 106.12.

[A] 104.12 Material and equipment reuse. Materials, equipment and devices shall not be reused unless such elements have been reconditioned, tested, placed in good and proper working condition and *approved*.

SECTION 105 PERMITS

[A] 105.1 When required. Any owner, or owner's authorized agent who desires to construct, enlarge, alter, repair, move, or demolish a pool or spa or to erect, install, enlarge, alter, repair, remove, convert or replace any system, the installation of which is regulated by this code, or to cause any such work to be performed, shall first make application to the code official and obtain the required permit for the work.

[A] 105.2 Application for permit. Each application for a permit, with the required fee, shall be filed with the code official on a form furnished for that purpose and shall contain a general description of the proposed work and its location. The application shall be signed by the owner or the owner's authorized agent. The permit application shall contain such other information required by the code official.

[A] 105.3 Construction documents. Construction documents, engineering calculations, diagrams and other such data shall be submitted in two or more sets with each application for a permit. The code official shall require construction documents, computations and specifications to be prepared and designed by a registered design professional where required by state law. Construction documents shall be drawn to scale and shall be of sufficient clarity to indicate the location, nature and extent of the work proposed and show in detail that the work conforms to the provisions of this code.

[A] 105.4 Time limitation of application. An application for a permit for any proposed work shall be deemed to have been abandoned 180 days after the date of filing unless such application has been pursued in good faith or a permit has been issued; except that the code official is authorized to grant one or more extensions of time for additional periods not exceeding 180 days each. The extension shall be requested in writing and justifiable cause demonstrated.

[A] 105.5 Permit issuance. The application, construction documents and other data filed by an applicant for permit shall be reviewed by the code official. If the code official finds that the proposed work conforms to the requirements of this code and laws and ordinances applicable thereto, and that the fees specified in Section 105.6 have been paid, a permit shall be issued to the applicant.

[A] 105.5.1 Approved construction documents. When the code official issues the permit where construction documents are required, the construction documents shall be endorsed in writing and stamped "APPROVED." Such *approved* construction documents shall not be changed, modified or altered without authorization from the code official. Work shall be done in accordance with the *approved* construction documents.

The code official shall have the authority to issue a permit for the construction of a part of a system before the entire construction documents for the whole system have been submitted or *approved*, provided that adequate information and detailed statements have been filed complying with pertinent requirements of this code. The holders of such permit shall proceed at their own risk without assurance that the permit for the entire system will be granted.

[A] 105.5.2 Validity. The issuance of a permit or approval of construction documents shall not be construed to be a permit for, or an approval of, any violation of any of the provisions of this code or any other ordinance of the jurisdiction. Any permit presuming to give authority to violate or cancel the provisions of this code shall not be valid.

The issuance of a permit based on construction documents and other data shall not prevent the code official from thereafter requiring the correction of errors in said construction documents and other data or from preventing building operations being carried on thereunder where in violation of this code or of other ordinances of this jurisdiction.

[A] 105.5.3 Expiration. Every permit issued shall become invalid unless the work authorized by such permit is commenced within 180 days after its issuance, or if the work authorized by such permit is suspended or abandoned for a period of 180 days after the time the work is commenced. The code official is authorized to grant, in writing, one or more extensions of time, for a period not more than 180 days. The extension shall be requested in writing and justifiable cause demonstrated.

[A] 105.5.4 Extensions. Any permittee holding an unexpired permit shall have the right to apply for an extension of the time within which the permittee will commence work under that permit when work is unable to be commenced within the time required by this section for good and satisfactory reasons. The code official shall extend the time for action by the permittee for a period not exceeding 180 days if there is reasonable cause. The fee for an extension shall be one-half the amount required for a new permit for such work.

[A] 105.5.5 Suspension or revocation of permit. The code official shall revoke a permit or approval issued under the provisions of this code in case of any false statement or misrepresentation of fact in the application or on the construction documents on which the permit or approval was based.

[A] 105.5.6 Retention of construction documents. One set of *approved* construction documents shall be retained by the code official for a period of not less than 180 days from date of completion of the permitted work, or as required by state or local laws. One set of *approved* construction documents shall be returned to the applicant, and said set shall be kept on the site of the building or work at all times during which the work authorized thereby is in progress.

[A] 105.6 Fees. A permit shall not be valid until the fees prescribed by law have been paid. An amendment to a permit shall not be released until the additional fee, if any, has been paid.

[A] 105.6.1 Work commencing before permit issuance. Any person who commences any work on a system before obtaining the necessary permits shall be subject to a fee as indicated in the adopted fee schedule and would be in addition to the required permit fees.

[A] 105.6.2 Fee schedule. The fees for work shall be as indicated in the following schedule:

[JURISDICTION TO INSERT APPROPRIATE SCHEDULE]

[A] 105.6.3 Fee refunds. The code official shall authorize the refunding of fees as follows:

1. The full amount of any fee paid hereunder that was erroneously paid or collected.
2. Not more than **[SPECIFY PERCENTAGE]** percent of the permit fee paid when no work has been done under a permit issued in accordance with this code.
3. Not more than **[SPECIFY PERCENTAGE]** percent of the plan review fee paid when an application for a permit for which a plan review fee has been paid is withdrawn or canceled before any plan review effort has been expended.

The code official shall not authorize the refunding of any fee paid except upon written application filed by the original permittee not later than 180 days after the date of fee payment.

SECTION 106 INSPECTIONS

[A] 106.1 General. Construction or work for which a permit is required shall be subject to inspection by the code official and such construction or work shall remain visible and able to be accessed for inspection purposes until *approved*. Approval as a result of an inspection shall not be construed to be an approval of a violation of the provisions of this code or of other ordinances of the jurisdiction. Inspections presuming to give authority to violate or cancel the provisions of this code or of other ordinances of the jurisdiction shall not be valid. It shall be the duty of the permit applicant to cause the work to remain accessible and exposed for inspection purposes. Neither the code official nor the jurisdiction shall be liable for expense entailed in the removal or replacement of any material required to allow inspection.

[A] 106.2 Preliminary inspection. Before issuing a permit, the code official is authorized to examine or cause to be examined buildings, structures and sites for which an application has been filed.

[A] 106.3 Required inspections and testing. Pool and spa installations or alterations thereto, including equipment, piping, and appliances related thereto, shall be inspected by the code official to ensure compliance with the requirements of this code.

[A] 106.4 Other inspections. In addition to the inspections specified in Sections 106.2 and 106.3, the code official is authorized to make or require other inspections of any construction work to ascertain compliance with the provisions of this code and other laws that are enforced.

[A] 106.5 Inspection request. It shall be the duty of the holder of the permit or their duly authorized agent to notify the code official when work is ready for inspection. It shall be the duty of the permit holder to provide access to and means for inspections of such work that are required by this code.

[A] 106.6 Approval required. Work shall not be done beyond the point indicated in each successive inspection without first obtaining the approval of the code official. The code official, upon notification, shall make the requested inspection and shall either indicate the portion of the construction that is satisfactory as completed, or notify the permit holder or his or her agent wherein the same fails to comply with this code. Any portions that do not comply shall be corrected and such portion shall not be covered or concealed until authorized by the code official.

[A] 106.7 Approved agencies. Test reports submitted to the code official for consideration shall be developed by *approved* agencies that have satisfied the requirements as to qualifications and reliability.

[A] 106.8 Evaluation and follow-up inspection services. Prior to the approval of a closed, prefabricated system and the issuance of a permit, the code official shall require the submittal of an evaluation report on each prefabricated system indicating the complete details of the system, including a description of the system and its components, the basis on which the system is being evaluated, test results and similar information, and other data as necessary for the code official to determine conformance to this code.

[A] 106.9 Evaluation service. The code official shall designate the evaluation service of an *approved* agency as the evaluation agency, and review such agency's evaluation report for adequacy and conformance to this code.

[A] 106.10 Follow-up inspection. Except where ready access is provided to systems, service equipment and accessories for complete inspection at the site without disassembly or dismantling, the code official shall conduct the frequency of in-plant inspections necessary to ensure conformance to the *approved* evaluation report or shall designate an independent, *approved* inspection agency to conduct such inspections. The inspection agency shall furnish the code official with the follow-up inspection manual and a report of inspections on request, and the system shall have an identifying label permanently affixed to the system indicating that factory inspections have been performed.

[A] 106.11 Test and inspection records. Required test and inspection records shall be available to the code official at all times during the fabrication of the system and the installation of the system, or such records as the code official designates shall be filed.

[A] 106.12 Special inspections. Special inspections of alternative engineered design systems shall be conducted in accordance with Sections 106.12.1 and 106.12.2.

[A] 106.12.1 Periodic inspection. The registered design professional or designated inspector shall periodically inspect and observe the alternative engineered design to determine that the installation is in accordance with the *approved* construction documents. Discrepancies shall be brought to the immediate attention of the contractor for correction. Records shall be kept of inspections.

[A] 106.12.2 Written report. The registered design professional shall submit a final report in writing to the code official upon completion of the installation, certifying that the alternative engineered design conforms to the *approved* construction documents. A notice of approval for the system shall not be issued until a written certification has been submitted.

[A] 106.13 Testing. Systems shall be tested as required by this code. Tests shall be made by the permit holder and the code official shall have the authority to witness such tests.

[A] 106.14 New, altered, extended or repaired systems. New systems and parts of existing systems that have been altered, extended or repaired shall be tested as prescribed by this code.

[A] 106.15 Equipment, material and labor for tests. Equipment, material and labor required for testing a system or part thereof shall be furnished by the permit holder.

[A] 106.16 Reinspection and testing. Where any work or installation does not pass any initial test or inspection, the necessary corrections shall be made to comply with this code. The work or installation shall then be resubmitted to the code official for inspection and testing.

[A] 106.17 Approval. After the prescribed tests and inspections indicate that the work complies in all respects with this code, a notice of approval shall be issued by the code official.

[A] 106.17.1 Revocation. The code official is authorized to, in writing, suspend or revoke a notice of approval issued under the provisions of this code wherever the notice is issued in error, or on the basis of the incorrect information supplied, or where it is determined that the building or structure, premise, system or portion thereof is in violation of any ordinance or regulation or any of the provisions of this code.

[A] 106.18 Temporary connection. The code official shall have the authority to authorize the temporary connection of the building or system to the utility source for the purpose of testing systems.

[A] 106.19 Connection of service utilities. A person shall not make connections from a utility, source of energy, fuel, power, water system or sewer system to any building or system that is regulated by this code for which a permit is required until authorized by the code official.

SECTION 107 VIOLATIONS

[A] 107.1 Unlawful acts. It shall be unlawful for any person, firm or corporation to erect, construct, alter, repair, remove, demolish or utilize any system, or cause same to be done, in conflict with or in violation of any of the provisions of this code.

[A] 107.2 Notice of violation. The code official shall serve a notice of violation or order to the person responsible for the erection, installation, alteration, extension, repair, removal or demolition of work in violation of the provisions of this code, or in violation of a detail statement or the *approved* construction documents there under, or in violation of a permit or certificate issued under the provisions of this code. Such order shall direct the discontinuance of the illegal action or condition and the abatement of the violation.

[A] 107.3 Prosecution of violation. If the notice of violation is not complied with promptly, the code official shall request the legal counsel of the jurisdiction to institute the appropriate proceeding at law or in equity to restrain, correct or abate such violation, or to require the removal or termination of the unlawful pool or spa in violation of the provisions of this code or of the order or direction made pursuant thereto.

[A] 107.4 Violation penalties. Any person who shall violate a provision of this code or shall fail to comply with any of the requirements thereof or who shall erect, install, alter or repair a pool or spa in violation of the *approved* construction documents or directive of the code official, or of a permit or certificate issued under the provisions of this code, shall be guilty of a [SPECIFY OFFENSE], punishable by a fine of not more than [AMOUNT] dollars or by imprisonment not exceeding [NUMBER OF DAYS], or both such fine and imprisonment. Each day that a violation continues after due notice has been served shall be deemed a separate offense.

[A] 107.5 Stop work orders. Upon notice from the *code official*, work on any system that is being performed contrary to the provisions of this code or in a dangerous or unsafe manner shall immediately cease. Such notice shall be in writing and shall be given to the owner of the property, or to the owner's authorized agent, or to the person performing the work. The notice shall state the conditions under which work is authorized to resume. Where an emergency exists, the code official shall not be required to give a written notice prior to stopping the work. Any person who shall continue any work in or about the structure after having been served with a stop work order, except such work as that person is directed to perform to remove a violation or unsafe condition, shall be liable to a fine of not less than [AMOUNT] dollars or more than [AMOUNT] dollars.

[A] 107.6 Abatement of violation. The imposition of the penalties herein prescribed shall not preclude the legal officer of the jurisdiction from instituting appropriate action to prevent violation, or to prevent illegal use of a pool or spa, or to stop an illegal act, conduct, business or utilization of the plumbing on or about any premises.

[A] 107.7 Unsafe systems. Any system regulated by this code that is unsafe or that constitutes a fire or health hazard, insanitary condition, or is otherwise dangerous to human life is hereby declared unsafe. Any use of a system regulated by this code constituting a hazard to safety, health or public welfare by reason of inadequate maintenance, dilapidation, obsolescence, fire hazard, disaster, damage or abandonment is hereby declared an unsafe use. Any such unsafe system is hereby declared to be a public nuisance and shall be abated by repair, rehabilitation, demolition or removal.

[A] 107.7.1 Authority to condemn a system. Where the code official determines that any system, or portion thereof, regulated by this code has become hazardous to life, health or property or has become insanitary, the code official shall order in writing that such system either be removed or restored to a safe or sanitary condition. A time limit for compliance with such order shall be specified in the written notice. A person shall not use or maintain a defective system after receiving such notice.

Where such a system is to be disconnected, written notice as prescribed in Section 107.2 shall be given. In cases of immediate danger to life or property, such disconnection shall be made immediately without such notice.

[A] 107.7.2 Authority to disconnect service utilities. The code official shall have the authority to authorize disconnection of utility service to the pool or spa regulated by the technical codes in case of an emergency, where necessary, to eliminate an immediate danger to life or property. Where possible, the owner or the owner's authorized agent and occupant of the building where the pool or spa is located shall be notified of the decision to disconnect utility service prior to taking such action. If not notified prior to disconnecting, the owner, the owner's authorized agent or the occupant of the building shall be notified in writing, as soon as practical thereafter.

[A] 107.7.3 Connection after order to disconnect. A person shall not make connections from any energy, fuel, power supply or water distribution system, or supply energy, fuel or water to any equipment regulated by this code that has been disconnected or ordered to be disconnected by the code official or the use of which has been ordered to be discontinued by the code official until the code official authorizes the reconnection and use of such equipment.

When any system is maintained in violation of this code, and in violation of any notice issued pursuant to the provisions of this section, the code official shall institute any appropriate action to prevent, restrain, correct or abate the violation.

SECTION 108
MEANS OF APPEAL

[A] 108.1 Application for appeal. Any person shall have the right to appeal a decision of the code official to the board of appeals. An application for appeal shall be based on a claim that the true intent of this code or the rules legally adopted there under have been incorrectly interpreted, the provisions of this code do not fully apply, or an equally good or better form of construction is proposed. The application shall be filed on a form obtained from the code official within 20 days after the notice was served.

[A] 108.2 Membership of board. The board of appeals shall consist of five members appointed by the chief appointing authority as follows: one for 5 years, one for 4 years, one for 3 years, one for 2 years and one for 1 year. Thereafter, each new member shall serve for 5 years or until a successor has been appointed.

[A] 108.2.1 Qualifications. The board of appeals shall consist of five individuals, one from each of the following professions or disciplines:

1. Registered design professional who is a registered architect; or a builder or superintendent of building construction with not less than 10 years' experience, 5 years of which shall have been in responsible charge of work.
2. Registered design professional with structural engineering or architectural experience.
3. Registered design professional with mechanical and plumbing engineering experience; or a mechanical and plumbing contractor with not less than 10 years' experience, 5 years of which shall have been in responsible charge of work.
4. Registered design professional with electrical engineering experience; or an electrical contractor with not less than 10 years' experience, 5 years of which shall have been in responsible charge of work.
5. Registered design professional with pool or spa experience; or a contractor with not less than 10 years' experience, 5 years of which shall have been in responsible charge of work.

[A] 108.2.2 Alternate members. The chief appointing authority shall appoint two alternate members who shall be called by the board chairman to hear appeals during the absence or disqualification of a member. Alternate members shall possess the qualifications required for board membership, and shall be appointed for 5 years or until a successor has been appointed.

[A] 108.2.3 Chairman. The board shall annually select one of its members to serve as chairman.

[A] 108.2.4 Disqualification of member. A member shall not hear an appeal in which that member has any personal, professional or financial interest.

[A] 108.2.5 Secretary. The chief administrative officer shall designate a qualified clerk to serve as secretary to the board. The secretary shall file a detailed record of proceedings in the office of the chief administrative officer.

[A] 108.2.6 Compensation of members. Compensation of members shall be determined by law.

[A] 108.3 Notice of meeting. The board shall meet upon notice from the chairman, within 10 days of the filing of an appeal or at stated periodic meetings.

[A] 108.4 Open hearing. Hearings before the board shall be open to the public. The appellant, the appellant's representative, the code official and any person whose interests are affected shall be given an opportunity to be heard.

[A] 108.4.1 Procedure. The board shall adopt and make available to the public through the secretary procedures under which a hearing will be conducted. The procedures shall not require compliance with strict rules of evidence, but shall mandate that only relevant information be received.

[A] 108.5 Postponed hearing. When five members are not present to hear an appeal, either the appellant or the appellant's representative shall have the right to request a postponement of the hearing.

[A] 108.6 Board decision. The board shall modify or reverse the decision of the code official by a concurring vote of three members.

[A] 108.6.1 Resolution. The decision of the board shall be by resolution. Certified copies shall be furnished to the appellant and to the code official.

[A] 108.6.2 Administration. The code official shall take immediate action in accordance with the decision of the board.

[A] 108.7 Court review. Any person, whether or not a previous party of the appeal, shall have the right to apply to the appropriate court to correct errors of law. Application for review shall be made in the manner and time required by law following the filing of the decision in the office of the chief administrative officer.

CHAPTER 2
DEFINITIONS

User note:

About this chapter: *Codes, by their very nature, are technical documents. Every word, term and punctuation mark can add to or change the meaning of a technical requirement. It is necessary to maintain a consensus on the specific meaning of each term contained in the code. Chapter 2 performs this function by stating clearly what specific terms mean for the purpose of the code.*

SECTION 201
GENERAL

201.1 Scope. Unless otherwise expressly stated, the following words and terms shall, for the purposes of this code, have the meanings shown in this chapter.

201.2 Interchangeability. Words used in the present tense include the future; words stated in the masculine gender include the feminine and neuter; the singular number includes the plural and the plural, the singular.

201.3 Terms defined in other codes. Where terms are not defined in this code and are defined in the *International Building Code*, *International Energy Conservation Code*, *International Fire Code*, *International Fuel Gas Code*, *International Mechanical Code*, *International Plumbing Code* or *International Residential Code*, such terms shall have the meanings ascribed to them as in those codes.

201.4 Terms not defined. Where terms are not defined through the methods authorized by this section, such terms shall have ordinarily accepted meanings such as the context implies.

SECTION 202
DEFINITIONS

ACCESSIBLE. Signifies access that requires the removal of an access panel or similar removable obstruction.

ACTIVITY POOL. A pool designed primarily for play activity that uses constructed features and devices including lily pad walks, flotation devices, small slide features, and similar attractions.

AIR INDUCTION SYSTEM. A system whereby a volume of air is introduced into hollow ducting built into a spa floor, bench, or hydrotherapy jets.

[A] ALTERATION. Any construction or renovation to an existing pool or spa other than repair.

[A] APPROVED. Acceptable to the code official or authority having jurisdiction.

[A] APPROVED AGENCY. An established and recognized agency regularly engaged in conducting tests or furnishing inspection services, or furnishing product certification where such agency has been *approved* by the code official.

AQUATIC RECREATION FACILITY. A facility that is designed for free-form aquatic play and recreation. The facilities may include, but are not limited to, wave or surf action pools, leisure rivers, sand bottom pools, vortex pools, activity pools, inner tube rides, body slides and interactive play attractions.

BACKWASH. The process of cleansing the filter medium or elements by the reverse flow of water through the filter.

BACKWASH CYCLE. The time required to backwash the filter medium or elements and to remove debris in the pool or spa filter.

BARRIER. A permanent fence, wall, building wall, or combination thereof that completely surrounds the pool or spa and obstructs the access to the pool or spa. The term "permanent" shall mean not being able to be removed, lifted, or relocated without the use of a tool.

BATHER. A person using a pool, spa or hot tub and adjoining deck area for the purpose of water sports, recreation, therapy or related activities.

BATHER LOAD. The number of persons in the pool or spa water at any given moment or during any stated period of time.

BEACH ENTRY. Sloping entry starting above the waterline at deck level and ending below the waterline. The presence of sand is not required. Also called "zero entry."

CHEMICAL FEEDER. A floating or mechanical device for adding a chemical to pool or spa water.

CIRCULATION EQUIPMENT. The components of a circulation system.

CIRCULATION SYSTEM. The mechanical components that are a part of a recirculation system on a pool or spa. Circulation equipment may be, but is not limited to, categories of pumps, hair and lint strainers, filters, valves, gauges, meters, heaters, surface skimmers, inlet fittings, outlet fittings and chemical feeding devices. The components have separate functions, but where connected to each other by piping, perform as a coordinated system for purposes of maintaining pool or spa water in a clear and sanitary condition.

[A] CODE OFFICIAL. The officer or other designated authority charged with the administration and enforcement of this code, or a duly authorized representative.

[A] CONSTRUCTION DOCUMENTS. Written, graphic and pictorial documents prepared or assembled for describing the design, location and physical characteristics of the elements of a project necessary for obtaining a building permit.

DECK. An area immediately adjacent to or attached to a pool or spa that is specifically constructed or installed for sitting, standing, or walking.

DEEP AREA. Water depth areas exceeding 5 feet (1524 mm).

DESIGN PROFESSIONAL. An individual who is registered or licensed to practice his or her respective design profession as defined by the statutory requirements of the professional registration or licensing laws of the state or jurisdiction in which the project is to be constructed.

DESIGN RATE OF FLOW. The rate of flow used for design calculations in a system.

DESIGN WATERLINE. The centerline of the *skimmer* or other point as defined by the designer of the pool or spa.

DIVING AREA. The area of a swimming pool that is designed for diving.

DIVING BOARD. A flexible board secured at one end that is used for diving such as a spring board or a jump board.

DIVING PLATFORM. A stationary platform designed for diving.

DIVING STAND. Any supporting device for a springboard, jump board or diving board.

EXERCISE SPA (Also known as a swim spa). Variants of a spa in which the design and construction includes specific features and equipment to produce a water flow intended to allow recreational physical activity including, but not limited to, swimming in place. Exercise spas can include peripheral jetted seats intended for water therapy, heater, circulation and filtration system, or can be a separate distinct portion of a combination spa/exercise spa and can have separate controls. These spas are of a design and size such that they have an unobstructed volume of water large enough to allow the 99th Percentile Man as specified in APSP 16 to swim or exercise in place.

EXISTING POOL OR SPA. A pool or spa constructed prior to the date of adoption of this code, or one for which a legal building permit has been issued.

FILTER. A device that removes undissolved particles from water by recirculating the water through a porous substance such as filter medium or elements.

FILTRATION. The process of removing undissolved particles from water by recirculating the water through a porous substance such as filter medium or elements.

[BS] FLOOD HAZARD AREA. The greater of the following two areas:

1. The area within a flood plain subject to a 1-percent or greater chance of flooding in any year.
2. The area designated as *a flood hazard area* on a community's flood hazard map, or otherwise legally designated.

FLUME. A trough-like or tubular structure, generally recognized as a water slide, that directs the path of travel and the rate of descent by the rider.

GUTTER. Overflow trough in the perimeter wall of a pool that is a component of the circulation system or flows to waste.

HAIR AND LINT STRAINER. A device attached on or in front of a pump to which the influent line (suction line) is connected for the purpose of entrapping lint, hair, or other debris that could damage the pump.

HANDHOLD. That portion of a pool or spa structure or a specific element that is at or above the *design waterline* that users in the pool grasp onto for support.

HANDRAIL. A support device that is intended to be gripped by a user for the purpose of resting or steadying, typically located within or at exits to the pool or spa or as part of a set of steps.

HYDROTHERAPY JET. A fitting that blends air and water, creating a high-velocity turbulent stream of air-enriched water.

JUMP BOARD. A manufactured diving board that has a coil spring, leaf spring, or comparable device located beneath the board that is activated by the force exerted by jumping on the board's end.

[A] JURISDICTION. The governmental unit that has adopted this code.

[A] LABEL. An identification applied on a product by the manufacturer that contains the name of the manufacturer, the function and performance characteristics of the product or material, and the name and identification of an *approved* agency and that indicates that the representative sample of the product or material has been tested and evaluated by an *approved* agency.

[A] LABELED. Equipment, materials or products to which has been affixed a label, seal, symbol or other identifying mark of a nationally recognized testing laboratory, *approved* agency or other organization concerned with product evaluation that maintains periodic inspection of the production of the above-labeled items and whose *labeling* indicates either that the equipment, material or product meets identified standards or has been tested and found suitable for a specified purpose.

LADDER. A structure for ingress and egress that usually consists of two long parallel side pieces joined at intervals by crosspieces such as treads.

Type A double access ladder. An "A-Frame" ladder that straddles the pool wall of an above-ground pool and provides ingress and egress and is intended to be removed when not in use.

Type B limited access ladder. An "A-Frame" ladder that straddles the pool wall of an above-ground/onground pool. Type B ladders are removable and have a built-in feature that prevents entry to the pool when the pool is not in use.

Type C ladder. A "ground to deck" staircase ladder that allows access to an above-ground pool deck and has a built-in entry-limiting feature.

Type D in-pool ladder. Located in the pool to provide a means of ingress and egress from the pool to the deck.

Type E or F in-pool staircase ladder. Located in the pool to provide a means of ingress and egress from the pool to the deck.

LIFELINE. An anchored line thrown to aid in rescue.

[A] LISTED. Equipment, materials, products or services included in a list published by an organization acceptable to the code official and concerned with evaluation of products or services that maintains periodic inspection of production of *listed* equipment or materials or periodic evaluation of services and whose listing states either that the equipment, material, product or service meets identified standards or has been tested and found suitable for a specified purpose.

MAINTAINED ILLUMINATION. The value, in foot-candles or equivalent units, below which the average illuminance on a specified surface is not allowed to fall. *Maintained illumination* equals the initial average illuminance on the specified surface with new lamps, multiplied by the light loss factor (LLF), to account for reduction in lamp intensity over time.

NEGATIVE EDGE. See "Vanishing edge."

NONENTRY AREA. An area of the deck from which entry into the pool or spa is prohibited.

ONGROUND STORABLE POOL. A pool that can be disassembled for storage or transport. This includes portable pools with flexible or nonrigid walls that achieve their structural integrity by means of uniform shape, a support frame or a combination thereof, and that can be disassembled for storage or relocation.

OVERFLOW GUTTER. The *gutter* around the top perimeter of the pool or spa, which is used to skim the surface.

[A] OWNER. Any person, agent, operator, entity, firm or corporation having any legal or equitable interest in the property; or recorded in the official records of the state, county or municipality as holding an interest or title to the property; or otherwise having possession or control of the property, including the guardian of the estate of any such person, and the executor or administrator of the estate of such person if ordered to take possession of real property by a court.

[A] PERMIT. An official document or certificate issued by the authority having jurisdiction that authorizes performance of a specified activity.

POOL. See "Public swimming pool" and "Residential swimming pool."

POWER SAFETY COVER. A pool cover that is placed over the water area, and is opened and closed with a motorized mechanism activated by a control switch.

PUBLIC SWIMMING POOL (Public Pool). A pool, other than a *residential* pool, that is intended to be used for swimming or bathing and is operated by an owner, lessee, operator, licensee or concessionaire, regardless of whether a fee is charged for use. Public pools shall be further classified and defined as follows:

Class A competition pool. A pool intended for use for accredited competitive aquatic events such as Federation Internationale De Natation (FINA), USA Swimming, USA Diving, USA Synchronized Swimming, USA Water Polo, National Collegiate Athletic Association (NCAA), or the National Federation of State High School Associations (NFHS).

Class B public pool. A pool intended for public recreational use that is not identified in the other classifications of public pools.

Class C semi-public pool. A pool operated solely for and in conjunction with lodgings such as hotels, motels, apartments or condominiums.

Class D-1 wave action pool. A pool designed to simulate breaking or cyclic waves for purposes of general play or surfing.

Class D-2 activity pool. A pool designed for casual water play ranging from simple splashing activity to the use of attractions placed in the pool for recreation.

Class D-3 catch pool. A body of water located at the termination of a manufactured waterslide attraction. The body of water is provided for the purpose of terminating the slide action and providing a means for exit to a deck or walkway area.

Class D-4 leisure river. A manufactured stream of water of near-constant depth in which the water is moved by pumps or other means of propulsion to provide a river-like flow that transports bathers over a defined path that may include water features and play devices.

Class D-5 vortex pool. A circular pool equipped with a method of transporting water in the pool for the purpose of propelling riders at speeds dictated by the velocity of the moving stream of water.

Class D-6 interactive play attraction. A manufactured water play device or a combination of water-based play devices in which water flow volumes, pressures or patterns can be varied by the bather without negatively influencing the hydraulic conditions for other connected devices. These attractions incorporate devices or activities such as slides, climbing and crawling structures, visual effects, user-actuated mechanical devices and other elements of bather-driven and bather-controlled play.

Class E. Pools used for instruction, play or therapy and with temperatures above 86°F (30°C).

Class F. Class F pools are wading pools and are covered within the scope of this code as set forth in Section 405.

Public pools are either a diving or nondiving type. Diving types of public pools are classified into types as an indication of the suitability of a pool for use with diving equipment.

Types VI–IX. Public pools suitable for the installation of diving equipment by type.

Type O. A nondiving public pool.

RECESSED TREADS. A series of vertically spaced cavities in a pool or spa wall creating tread areas for step holes.

RECIRCULATION SYSTEM. See "Circulation system."

[A] REPAIR. The reconstruction or renewal of any part of a pool or spa for the purpose of its maintenance or to correct damage.

RESIDENTIAL. For purposes of this code, *residential* applies to detached one- and two-family dwellings and townhouses not more than three stories in height.

RESIDENTIAL SWIMMING POOL (Residential Pool). A pool intended for use that is accessory to a *residential* setting and available only to the household and its guests. Other pools shall be considered to be public pools for purposes of this code.

Types I–V. *Residential* pools suitable for the installation of diving equipment by type.

Type O. A nondiving *residential* pool.

RETURN INLET. The aperture or fitting through which the water under positive pressure returns into a pool.

RING BUOY. A ring-shaped floating buoy capable of supporting a user, usually attached to a throwing line.

ROPE AND FLOAT LINE. A continuous line not less than $^1/_4$ inch (6 mm) in diameter that is supported by buoys and attached to opposite sides of a pool to separate the deep and shallow ends.

RUNOUT. A continuation of water slide flume surface where riders are intended to decelerate and come to a stop.

SAFETY COVER. A structure, fabric or assembly, along with attendant appurtenances and anchoring mechanisms, that is temporarily placed or installed over an entire pool, spa or hot tub and secured in place after all bathers are absent from the water.

SHALL. The term, where used in the code, is construed as mandatory.

SHALLOW AREAS. Portions of a pool or spa with water depths less than 5 feet (1524 mm).

SKIMMER. A device installed in the pool or spa that permits the removal of floating debris and surface water to the filter.

SLIP RESISTANT. A surface that has been treated or constructed to significantly reduce the chance of a user slipping. The surface shall not be an abrasion hazard.

SLOPE BREAK. Occurs at the point where the slope of the pool floor changes to a greater slope.

SPA. A product intended for the immersion of persons in temperature-controlled water circulated in a closed system, and not intended to be drained and filled with each use. A spa usually includes a filter, an electric, solar or gas heater, a pump or pumps, and a control, and can include other equipment, such as lights, blowers, and water-sanitizing equipment.

Permanent residential spa. A spa, intended for use that is accessory to a *residential* setting and available to the household and its guests and where the water heating and water-circulating equipment is not an integral part of the product. The spa is intended as a permanent plumbing fixture and not intended to be moved.

Portable residential spa. A spa intended for use that is accessory to a *residential* setting and available to the household and its guests and where it is either self- contained or nonself-contained.

Public spa. A spa other than a permanent *residential* spa or portable *residential* spa that is intended to be used for bathing and is operated by an owner, licensee or concessionaire, regardless of whether a fee is charged for use.

Self-contained spa. A factory-built spa in which all control, water heating and water-circulating equipment is an integral part of the product. Self-contained spas may be permanently wired or cord connected.

Nonself-contained spa. A factory-built *spa* in which the water heating and circulating equipment is not an integral part of the product. Nonself-contained spas may employ separate components such as an individual filter, pump, heater and controls, or they can employ assembled combinations of various components.

SPRAY POOL. A pool or basin occupied by construction features that spray water in various arrays for the purpose of wetting the persons playing in the spray streams.

SUBMERGED VACUUM FITTING. A fitting intended to provide a point of connection for suction side automatic swimming pool, *spa*, and hot tub cleaners.

SUCTION OUTLET. A submerged fitting, fitting assembly, cover/grate and related components that provide a localized low-pressure area for the transfer of water from a swimming pool, spa or hot tub. Submerged suction outlets have been referred to as main drains.

SURFACE SKIMMING SYSTEM. A device or system installed in the pool or spa that permits the removal of floating debris and surface water to the filter.

SURGE CAPACITY. The storage volume in a surge tank, *gutter*, and plumbing lines.

SURGE TANK. A storage vessel within the pool recirculating system used to contain the water displaced by bathers.

SWIMOUT. An underwater seat area that is placed completely outside of the perimeter shape of the pool. Where located at the deep end, swimouts are permitted to be used as the deep-end means of entry or exit to the pool.

TUBE RIDE. A gravity flow attraction found at a waterpark designed to convey riders on an inner-tube-like device through a series of chutes, channels, flumes or pools.

TURNOVER RATE. The period of time, usually in hours, required to circulate a volume of water equal to the pool or spa capacity.

UNDERWATER LEDGE. A narrow shelf projecting from the side of a vertical structure whose dimensions are defined in the appropriate standard.

UNDERWATER SEAT. An underwater ledge that is placed completely inside the perimeter shape of the pool, generally located in the shallow end of the pool.

VANISHING EDGE. Water-feature detail in which water flows over the edge of not fewer than one of the pool walls and is collected in a catch basin. Also called "Negative edge."

WATERLINE. See "Design waterline."

WAVE POOL CAISSON. A large chamber used in wave generation. This chamber houses pulsing water and air surges in the wave generation process and is not meant for human occupancy.

ZERO ENTRY. See "Beach entry."

CHAPTER 3

GENERAL COMPLIANCE

User note:

About this chapter: *Chapter 3 covers general regulations for pool and spa installations. As many of these requirements would need to be repeated in Chapters 3 through 10, placing such requirements in only one location eliminates code development coordination issues with the same requirement in multiple locations. These general requirements can be superseded by more specific requirements for certain applications in Chapters 3 through 10.*

SECTION 301 GENERAL

301.1 Scope. The provisions of this chapter shall govern the general design and construction of public and *residential* pools and spas and related piping, equipment, and materials. Provisions that are unique to a specific type of pool or spa are located in Chapters 4 through 10.

301.1.1 Application of Chapters 4 through 10. Where differences occur between the provisions of this chapter and the provisions of Chapters 4 through 10, the provisions of Chapters 4 through 10 shall apply.

SECTION 302 ELECTRICAL, PLUMBING, MECHANICAL AND FUEL GAS REQUIREMENTS

302.1 Electrical. Electrical requirements for aquatic facilities shall be in accordance with NFPA 70 or the *International Residential Code*, as applicable in accordance with Section 102.7.1.

Exception: Internal wiring for portable *residential* spas and portable *residential* exercise spas.

302.2 Water service and drainage. Piping and fittings used for water service, makeup and drainage piping for pools and spas shall comply with the *International Plumbing Code*. Fittings shall be *approved* for installation with the piping installed.

302.3 Pipe, fittings and components. Pipe, fittings and components shall be *listed* and *labeled* in accordance with NSF 50 or NSF 14. Plastic jets, fittings, and outlets used in public spas shall be *listed* and *labeled* in accordance with NSF 50.

Exceptions:

1. Portable *residential* spas and portable *residential* exercise spas *listed* and *labeled* in accordance with UL 1563 or CSA C22.2 No. 218.1.
2. *Onground storable pools* supplied by the pool manufacturer as a kit that includes all pipe, fittings and components.

302.4 Concealed piping inspection. Piping, including process piping, that is installed in trenches, shall be inspected prior to backfilling.

302.5 Backflow protection. Water supplies for pools and spas shall be protected against backflow in accordance with the *International Plumbing Code* or the *International Residential Code*, as applicable in accordance with Section 102.7.1.

302.6 Wastewater discharge. Where wastewater from pools or spas, such as backwash water from filters and water from deck drains discharge to a building drainage system, the connection shall be through an air gap in accordance with the *International Plumbing Code* or the *International Residential Code* as applicable in accordance with Section 102.7.1.

302.7 Tests. Tests on water piping systems constructed of plastic piping shall not use compressed air for the test.

302.8 Maintenance. Pools and spas shall be maintained in a clean and sanitary condition, and in good repair.

302.8.1 Manuals. An operating and maintenance manual in accordance with industry-accepted standards shall be provided for each piece of equipment requiring maintenance.

SECTION 303 ENERGY

303.1 Energy consumption of pools and permanent spas. The energy consumption of pools and permanent spas shall be controlled by the requirements in Sections 303.1.1 through 303.1.3.

303.1.1 Heaters. The electric power to heaters shall be controlled by a readily accessible on-off switch that is an integral part of the heater, mounted on the exterior of the heater or external to and within 3 feet (914 mm) of the heater. Operation of such switch shall not change the setting of the heater thermostat. Such switches shall be in addition to a circuit breaker for the power to the heater. Gas-fired heaters shall not be equipped with continuously burning ignition pilots.

303.1.2 Time switches. Time switches or other control methods that can automatically turn off and on heaters and pump motors according to a preset schedule shall be installed for heaters and pump motors. Heaters and pump motors that have built-in time switches shall be in compliance with this section.

Exceptions:

1. Where public health standards require 24-hour pump operation.
2. Pumps that operate solar- or waste-heat recovery pool heating systems.

303.1.3 Covers. Outdoor heated pools and outdoor permanent spas shall be provided with a vapor-retardant cover or other *approved* vapor-retardant means in accordance with Section 104.11.

Exception: Where more than 70 percent of the energy for heating, computed over an operating season, is from a heat pump or solar energy source, covers or other vapor-retardant means shall not be required.

303.2 Portable spas. The energy consumption of electric-powered portable spas shall be controlled by the requirements of APSP 14.

303.3 Residential pools and permanent residential spas. The energy consumption of *residential* swimming pools and permanent *residential* spas shall be controlled in accordance with the requirements of APSP 15.

SECTION 304
FLOOD HAZARD AREAS

304.1 General. The provisions of Section 304 shall control the design and construction of pools and spas installed in *flood hazard areas.*

[BS] 304.2 Determination of impacts based on location. Pools and spas located in *flood hazard areas* indicated within the *International Building Code* or the *International Residential Code* shall comply with Section 304.2.1 or 304.2.2.

Exception: Pools and spas located in riverine *flood hazard areas* that are outside of designated floodways and pools and spas located in *flood hazard areas* where the source of flooding is tides, storm surges or coastal storms.

[BS] 304.2.1 Pools and spas located in designated floodways. Where pools and spas are located in designated floodways, documentation shall be submitted to the code official that demonstrates that the construction of the pools and spas will not increase the design flood elevation at any point within the jurisdiction.

[BS] 304.2.2 Pools and spas located where floodways have not been designated. Where pools and spas are located where design flood elevations are specified but floodways have not been designated, the applicant shall provide a floodway analysis that demonstrates that the proposed pool or spa and any associated grading and filling, will not increase the design flood elevation more than 1 foot (305 mm) at any point within the jurisdiction.

[BS] 304.3 Pools and spas in coastal high-hazard areas. Pools and spas installed in coastal high-hazard areas shall be designed and constructed in accordance with ASCE 24.

[BS] 304.4 Protection of equipment. Equipment shall be elevated to or above the design flood elevation or be anchored to prevent flotation and protected to prevent water from entering or accumulating within the components during conditions of flooding.

304.5 GFCI protection. Electrical equipment installed below the design flood elevation shall be supplied by branch circuits that have ground-fault circuit interrupter protection for personnel.

SECTION 305
BARRIER REQUIREMENTS

305.1 General. The provisions of this section shall apply to the design of barriers for restricting entry into areas having pools and spas. Where spas or hot tubs are equipped with a lockable safety cover complying with ASTM F1346 and swimming pools are equipped with a powered safety cover that complies with ASTM F1346, the areas where those spas, hot tubs or pools are located shall not be required to comply with Sections 305.2 through 305.7.

305.2 Outdoor swimming pools and spas. Outdoor pools and spas and indoor swimming pools shall be surrounded by a barrier that complies with Sections 305.2.1 through 305.7.

305.2.1 Barrier height and clearances. Barrier heights and clearances shall be in accordance with all of the following:

1. The top of the barrier shall be not less than 48 inches (1219 mm) above grade where measured on the side of the barrier that faces away from the pool or spa. Such height shall exist around the entire perimeter of the barrier and for a distance of 3 feet (914 mm) measured horizontally from the outside of the required barrier.
2. The vertical clearance between grade and the bottom of the barrier shall not exceed 2 inches (51 mm) for grade surfaces that are not solid, such as grass or gravel, where measured on the side of the barrier that faces away from the pool or spa.
3. The vertical clearance between a surface below the barrier to a solid surface, such as concrete, and the bottom of the required barrier shall not exceed 4 inches (102 mm) where measured on the side of the required barrier that faces away from the pool or spa.
4. Where the top of the pool or spa structure is above grade, the barrier shall be installed on grade or shall be mounted on top of the pool or spa structure. Where the barrier is mounted on the top of the pool or spa, the vertical clearance between the top of the pool or spa and the bottom of the barrier shall not exceed 4 inches (102 mm).

305.2.2 Openings. Openings in the barrier shall not allow passage of a 4-inch-diameter (102 mm) sphere.

305.2.3 Solid barrier surfaces. Solid barriers that do not have openings shall not contain indentations or protrusions that form handholds and footholds, except for normal construction tolerances and tooled masonry joints.

305.2.4 Mesh fence as a barrier. Mesh fences, other than chain link fences in accordance with Section 305.2.7, shall be installed in accordance with the manufacturer's instructions and shall comply with the following:

1. The bottom of the mesh fence shall be not more than 1 inch (25 mm) above the deck or installed surface or grade.
2. The maximum vertical clearance from the bottom of the mesh fence and the solid surface shall not permit

the fence to be lifted more than 4 inches (102 mm) from grade or decking.

3. The fence shall be designed and constructed so that it does not allow passage of a 4-inch (102 mm) sphere under any mesh panel. The maximum vertical clearance from the bottom of the mesh fence and the solid surface shall be not greater than 4 inches (102 mm) from grade or decking.
4. An attachment device shall attach each barrier section at a height not lower than 45 inches (1143 mm) above grade. Common attachment devices include, but are not limited to, devices that provide the security equal to or greater than that of a hook-and-eye-type latch incorporating a spring-actuated retaining lever such as a safety gate hook.
5. Where a hinged gate is used with a mesh fence, the gate shall comply with Section 305.3.
6. Patio deck sleeves such as vertical post receptacles that are placed inside the patio surface shall be of a nonconductive material.
7. Mesh fences shall not be installed on top of onground *residential* pools.

305.2.5 Closely spaced horizontal members. Where the barrier is composed of horizontal and vertical members and the distance between the tops of the horizontal members is less than 45 inches (1143 mm), the horizontal members shall be located on the pool or spa side of the fence. Spacing between vertical members shall not exceed $1^{3}/_{4}$ inches (44 mm) in width. Where there are decorative cutouts within vertical members, spacing within the cutouts shall not exceed $1^{3}/_{4}$ inches (44 mm) in width.

305.2.6 Widely spaced horizontal members. Where the barrier is composed of horizontal and vertical members and the distance between the tops of the horizontal members is 45 inches (1143 mm) or more, spacing between vertical members shall not exceed 4 inches (102 mm). Where there are decorative cutouts within vertical members, the interior width of the cutouts shall not exceed $1^{3}/_{4}$ inches (44 mm).

305.2.7 Chain link dimensions. The maximum opening formed by a chain link fence shall be not more than $1^{3}/_{4}$ inches (44 mm). Where the fence is provided with slats fastened at the top and bottom that reduce the openings, such openings shall be not greater than $1^{3}/_{4}$ inches (44 mm).

305.2.8 Diagonal members. Where the barrier is composed of diagonal members, the maximum opening formed by the diagonal members shall be not greater than $1^{3}/_{4}$ inches (44 mm). The angle of diagonal members shall be not greater than 45 degrees (0.79 rad) from vertical.

305.2.9 Clear zone. There shall be a clear zone of not less than 36 inches (914 mm) between the exterior of the barrier and any permanent structures or equipment such as pumps, filters and heaters that can be used to climb the barrier.

305.2.10 Poolside barrier setbacks. The pool or spa side of the required barrier shall be not less than 20 inches (508 mm) from the water's edge.

305.3 Gates. Access gates shall comply with the requirements of Sections 305.3.1 through 305.3.3 and shall be equipped to accommodate a locking device. Pedestrian access gates shall open outward away from the pool or spa, shall be self-closing and shall have a self-latching device.

305.3.1 Utility or service gates. Gates not intended for pedestrian use, such as utility or service gates, shall remain locked when not in use.

305.3.2 Double or multiple gates. Double gates or multiple gates shall have not fewer than one leaf secured in place and the adjacent leaf shall be secured with a self-latching device. The gate and barrier shall not have openings larger than $^{1}/_{2}$ inch (12.7 mm) within 18 inches (457 mm) of the latch release mechanism. The self-latching device shall comply with the requirements of Section 305.3.3.

305.3.3 Latches. Where the release mechanism of the self-latching device is located less than 54 inches (1372 mm) from grade, the release mechanism shall be located on the pool or spa side of the gate not less than 3 inches (76 mm) below the top of the gate, and the gate and barrier shall not have openings greater than $^{1}/_{2}$ inch (12.7 mm) within 18 inches (457 mm) of the release mechanism.

305.4 Structure wall as a barrier. Where a wall of a dwelling or structure serves as part of the barrier and where doors or windows provide direct access to the pool or spa through that wall, one of the following shall be required:

1. Operable windows having a sill height of less than 48 inches (1219 mm) above the indoor finished floor and doors shall have an alarm that produces an audible warning when the window, door or their screens are opened. The alarm shall be *listed* and *labeled* as a water hazard entrance alarm in accordance with UL 2017. In dwellings or structures not required to be Accessible units, Type A units or Type B units, the operable parts of the alarm deactivation switches shall be located 54 inches (1372 mm) or more above the finished floor. In dwellings or structures required to be Accessible units, Type A units or Type B units, the operable parts of the alarm deactivation switches shall be located not greater than 54 inches (1372 mm) and not less than 48 inches (1219 mm) above the finished floor.
2. A *safety cover* that is *listed* and *labeled* in accordance with ASTM F1346 is installed for the pools and spas.
3. An *approved* means of protection, such as self-closing doors with self-latching devices, is provided. Such means of protection shall provide a degree of protection that is not less than the protection afforded by Item 1 or 2.

305.5 Onground residential pool structure as a barrier. An onground *residential* pool wall structure or a barrier mounted on top of an onground *residential* pool wall structure shall serve as a barrier where all of the following conditions are present:

1. Where only the pool wall serves as the barrier, the bottom of the wall is on grade, the top of the wall is not less than 48 inches (1219 mm) above grade for the

entire perimeter of the pool, the wall complies with the requirements of Section 305.2 and the pool manufacturer allows the wall to serve as a barrier.

2. Where a barrier is mounted on top of the pool wall, the top of the barrier is not less than 48 inches (1219 mm) above grade for the entire perimeter of the pool, and the wall and the barrier on top of the wall comply with the requirements of Section 305.2.
3. Ladders or steps used as means of access to the pool are capable of being secured, locked or removed to prevent access except where the ladder or steps are surrounded by a barrier that meets the requirements of Section 305.
4. Openings created by the securing, locking or removal of ladders and steps do not allow the passage of a 4-inch (102 mm) diameter sphere.
5. Barriers that are mounted on top of onground *residential* pool walls are installed in accordance with the pool manufacturer's instructions.

305.6 Natural barriers. In the case where the pool or spa area abuts the edge of a lake or other natural body of water, public access is not permitted or allowed along the shoreline, and required barriers extend to and beyond the water's edge not less than 18 inches (457 mm), a barrier is not required between the natural body of water shoreline and the pool or spa.

305.7 Natural topography. Natural topography that prevents direct access to the pool or spa area shall include but not be limited to mountains and natural rock formations. A natural barrier approved by the governing body shall be acceptable provided that the degree of protection is not less than the protection afforded by the requirements of Sections 305.2 through 305.5.

SECTION 306
DECKS

306.1 General. The structural design and installation of decks around pools and spas shall be in accordance with the *International Residential Code* or the *International Building Code,* as applicable in accordance with Section 102.7 and this section.

306.2 Slip resistant. Decks, ramps, coping, and similar step surfaces shall be slip resistant and cleanable. Special features in or on decks such as markers, brand insignias, and similar materials shall be slip resistant.

306.3 Step risers and treads. Step risers for decks of public pools and spas shall be uniform and have a height not less than $3^3/_4$ inches (95 mm) and not greater than $7^1/_2$ inches (191 mm). The tread distance from front to back shall be not less than 11 inches (279 mm). Step risers for decks of *residential* pools and spas shall be uniform and shall have a height not exceeding $7^1/_2$ inches (191 mm). The tread distance from front to back shall be not less than 10 inches (254 mm).

306.4 Deck steps handrail required. Public pool and spa deck steps having three or more risers shall be provided with a handrail.

306.5 Slope. The minimum slope of decks shall be in accordance with Table 306.5 except where an alternative drainage method is provided that prevents the accumulation or pooling of water. The slope for decks, other than wood decks, shall be not greater than $^1/_2$ inch per foot (1 mm per 24 mm) except for ramps. The slope for wood and wood/plastic composite decks shall be not greater than $^1/_4$ inch per 1 foot (1 mm per 48 mm). Decks shall be sloped so that standing water will not be deeper than $^1/_8$ inch (3.2 mm), 20 minutes after the cessation of the addition of water to the deck.

306.6 Gaps. Gaps shall be provided between deck boards in wood and wood/plastic composite decks. Gaps shall be consistent with *approved* engineering methods with respect to the type of wood used and shall not cause a tripping hazard.

306.6.1 Maximum gap. The open gap between pool decks and adjoining decks or walkways, including joint material, shall be not greater than $^3/_4$ inch (19.1 mm). The difference in vertical elevation between the pool deck and the adjoining sidewalk shall be not greater than $^1/_4$ inch (6.4 mm).

306.7 Concrete joints. Isolation joints that occur where the pool coping meets the concrete deck shall be water tight.

306.7.1 Joints at coping. Joints that occur where the pool coping meets the concrete deck shall be installed to protect the coping and its mortar bed from damage as a result of the anticipated movement of adjoining deck.

306.7.2 Crack control. Joints in a deck shall be provided to minimize visible cracks outside of the control joints caused by imposed stresses or movement of the slab.

TABLE 306.5
MINIMUM DRAINAGE SLOPES FOR DECK SURFACES

SURFACE	MINIMUM DRAINAGE SLOPE (INCH PER FOOT)
Carpet	$^1/_2$
Exposed aggregate	$^1/_4$
Textured, hand-finished concrete	$^1/_8$
Travertine/brick-set pavers, public pools or spas	$^3/_8$
Travertine/brick-set pavers, residential pools or spas	$^1/_8$
Wood	$^1/_8$
Wood/plastic composite	$^1/_8$

For SI: 1 inch = 25.4 mm, 1 foot = 304.8 mm.

306.7.3 Movement control. Areas where decks join existing concrete work shall be provided with a joint to protect the pool from damage caused by relative movement.

306.8 Deck edges. The edges of decks shall be radiused, tapered, or otherwise designed to eliminate sharp corners.

306.9 Valves under decks. Valves installed in or under decks shall be accessible for operation, service, and maintenance. Where access through the deck walking surface is required, an access cover shall be provided for the opening in the deck. Such access covers shall be slip resistant and secured.

306.9.1 Hose bibbs. Hose bibbs shall be provided for rinsing down the entire deck and shall be installed in accordance with the *International Plumbing Code* or *International Residential Code*, as applicable in accordance with Section 102.7.1, and shall be located not greater than 150 feet (45 720 mm) apart. Water-powered devices, such as water-powered lifts, shall have a dedicated hose bibb water source.

Exception: *Residential* pools and spas shall not be required to have hose bibbs located at 150-foot (45 720 mm) intervals, or have a dedicated hose bibb for water-powered devices.

SECTION 307 GENERAL DESIGN

307.1 General design requirements. Sections 307.1.1 through 307.1.4 shall apply to all pools and spas.

307.1.1 Glazing in hazardous locations. Hazardous locations for glazing shall be as defined in the *International Building Code* or the *International Residential Code*, as applicable in accordance with Section 102.7.1 of this code. Where glazing is determined to be in a hazardous location, the requirements for the glazing shall be in accordance with those codes, as applicable.

307.1.2 Colors and finishes. For other than *residential* pools and *residential* spas, the colors, patterns, or finishes of the pool and spa interiors shall not obscure objects or surfaces within the pool or spa.

307.1.3 Roofs or canopies. Roofs or canopies over pools and spas shall be in accordance with the *International Building Code* or *International Residential Code*, as applicable in accordance with Section 102.7.1 and shall be constructed so as to prevent water runoff into the pool or spa.

307.1.4 Accessibility. An accessible route to public pools and spas shall be provided in accordance with the *International Building Code*. Accessibility within public pools and spas shall be provided as required by the accessible recreational facilities provisions of the *International Building Code*.

307.2 Specific design and material requirements. Sections 307.2.1 through 307.2.4 shall apply to all pools and spas except for *listed* and *labeled* portable *residential* spas, and *listed* and *labeled* portable *residential* exercise spas.

307.2.1 Materials. Pools and spas and appurtenances thereto shall be constructed of materials that are nontoxic to humans and the environment; that are generally or commonly regarded to be impervious and enduring; that will withstand the design stresses; and that will provide a watertight structure with a smooth and easily cleanable surface without cracks or joints, excluding structural joints, or that will provide a watertight structure to which a smooth, easily cleaned surface/finish is applied or attached. Material surfaces that come in contact with the user shall be finished, so that they do not constitute a cutting, pinching, puncturing or abrasion hazard under casual contact and intended use.

307.2.1.1 Beach pools. Clean sand or similar material, where used in a beach pool environment, shall be used over an impervious surface. The sand area shall be designed and controlled so that the circulation system, maintenance, safety, sanitation, and operation of the pool are not adversely affected.

307.2.1.2 Compatibility. Assemblies of different materials shall be chemically and mechanically compatible for their intended use and environment.

307.2.2 Materials and structural design. Pools and spas shall conform to one or more of the standards indicated in Table 307.2.2. The structural design of pools and spas shall be in accordance with the *International Building Code* or the *International Residential Code*, as applicable in accordance with Section 102.7.1 of this code.

TABLE 307.2.2 RESERVOIRS AND SHELLS

MATERIAL	STANDARD
Fiberglass reinforced plastic	IAPMO Z124.7
Plastic	IAPMO Z124.7
Stainless steel (Types 316, 316L, 304, 304L)	ASTM A240
Tile	ANSI A108/A118/A136.1
Vinyl	ASTM D1593

307.2.2.1 Installation. Equipment for pools and spas shall be supported to prevent damage from misalignment and settling and located so as to allow access for inspection, servicing, removal and repair of component parts.

307.2.3 Freeze protection. In climates subject to freezing temperatures, outdoor pool and spa shells and appurtenances, piping, filter systems, pumps and motors, and other components shall be designed and constructed to provide protection from damage from freezing.

307.2.4 Surface condition. The surfaces within public pools and spas intended to provide footing for users shall be slip resistant and shall not cause injury during normal use.

SECTION 308 DIMENSIONAL DESIGN

308.1 Floor slope. The slope of the floor from the point of the first slope change to the deep area shall not exceed one unit vertical in three units horizontal (33-percent slope).

Exception: Portable *residential* spas and portable *residential* exercise spas.

308.2 Walls. Walls shall intersect with the floor at an angle or a transition profile. Where a transitional profile is provided at water depths of 3 feet (914 mm) or less, a transitional radius shall not exceed 6 inches (152 mm) and shall be tangent to the wall and is permitted to be tangent to or intersect the floor.

Exceptions:

1. Portable *residential* spas and portable *residential* exercise spas.
2. *Onground storable pools.*

308.3 Shape. This code is not intended to regulate the shape of a pool or spa other than to take into account the effect that a given shape will have on the safety of the occupants and to maintain the minimum required level of circulation to ensure sanitation.

308.4 Waterline. The *design waterline* shall have a maximum construction tolerance at the time of completion of the work of plus or minus $^1/_4$ inch (6.4 mm) for pools and spas with adjustable weir surface skimming systems, and plus or minus $^1/_8$ inch (3.2 mm) for pools and spas with nonadjustable surface skimming systems.

SECTION 309 EQUIPMENT

309.1 Electrically operated equipment. Electrically operated equipment shall be *listed* and *labeled* in accordance with applicable product standards.

Exception: Portable *residential* spas and portable *residential* exercise spas *listed* and *labeled* in accordance with UL 1563 or CSA C22.2 No. 218.1.

309.2 Treatment and circulation system equipment. Treatment and circulation system equipment for public pools and spas shall be *listed* and *labeled* in accordance with NSF 50 and other applicable standards.

SECTION 310 SUCTION ENTRAPMENT AVOIDANCE

310.1 General. Suction entrapment avoidance for pools and spas shall be provided in accordance with APSP 7.

Exceptions:

1. Portable spas and portable exercise spas *listed* and *labeled* in accordance with UL 1563 or CSA C22.2 No. 218.1.
2. Suction entrapment avoidance for wading pools shall be provided in accordance with Section 405.

SECTION 311 CIRCULATION SYSTEMS

311.1 General. The provisions of this section shall apply to circulation systems for pools and spas.

Exceptions:

1. Portable *residential* spas and portable *residential* exercise spas.
2. *Onground storable pools* supplied by the pool manufacturer as a kit that includes circulation system equipment that is in accordance with Section 704.

311.2 System design. A circulation system consisting of pumps, piping, return inlets and outlets, filters, and other necessary equipment shall be provided for the complete circulation of water. Wading pools and spas shall have separate dedicated filtering systems.

Exception: Separate filtering systems are not required for *residential* pools and spas.

311.2.1 Turnover rate. The equipment shall be sized to turn over the volume of water that the pool or spa is capable of containing as specified in this code for the specific installation.

311.2.2 Servicing. Circulation system components that require replacement or servicing shall be provided with access for inspection, repair, or replacement and shall be installed in accordance with the manufacturer's specifications.

311.2.3 Equipment anchorage. Pool and spa equipment and related piping shall be designed and installed in accordance with the manufacturer's instructions.

311.3 Water velocity. The water velocity in return lines shall not exceed 8 feet (2.4 m) per second. The water velocity in suction piping shall be as required by Section 310.

311.4 Piping and fittings. Plastic pipe and fittings used in circulation systems shall be nontoxic and shall be able to withstand the design operating pressures and conditions of the pool or spa. Plastic pipe shall be *listed* and *labeled* as complying with NSF 14. Circulation system piping shall be *listed* and *labeled* as complying with one of the standards in Table 311.4.

TABLE 311.4 CIRCULATION SYSTEM PIPE MATERIAL STANDARD

MATERIAL	STANDARD
Acrylonitrile butadiene styrene (ABS) plastic pipe	ASTM D1527
Chlorinated polyvinyl chloride (CPVC) plastic pipe and tubing	ASTM D2846; CSA B137.6
Copper or copper-alloy tubing	ASTM B88; ASTM B447
Polyvinyl chloride (PVC) hose	ASTM D1785; ASTM D2241; ASTM D2672; CSA B137.3
Polyvinyl chloride (PVC) plastic pipe	ASTM D1785; CSA B137.3
Stainless steel pipe, Types 304, 304L, 316, 316L	ASTM A312

311.4.1 Fittings. Fittings used in circulation systems shall be *listed* and *labeled* as complying with one of the standards in Table 311.4.1.

Exceptions:

1. Suction outlet fitting assemblies and manufacturer-provided components certified in accordance with APSP 16.
2. Skimmers and manufacturer-provided components.
3. *Gutter* overflow grates and fittings installed above or outside of the overflow point of the pool or spa.

311.4.2 Joints. Joints shall be made in accordance with manufacturer's instructions.

311.4.3 Piping subject to freezing. Piping subject to damage by freezing shall have a uniform slope in one direction and shall be equipped with valves for drainage or shall be capable of being evacuated to remove the water.

311.4.4 Suction outlet fitting assemblies. Suction outlet fitting assemblies shall be *listed* and *labeled* in compliance with APSP 16.

311.5 System draining. Equipment shall be designed and fabricated to drain the water from the equipment, together with exposed face piping, by removal of drain plugs, manipulating valves, or by other methods. Drainage shall be in accordance with manufacturer's specifications.

311.6 Pressure or vacuum gauge. Gauges shall be provided on the circulation system for public pools. Gauges shall be provided with ready access.

1. A pressure gauge shall be located downstream of the pump and between the pump and filter.
2. A vacuum gauge shall be located between the pump and filter and upstream of the pump.

311.7 Flow measurement. Public swimming pools and wading pools shall be equipped with a flow-measuring device that indicates the rate of flow through the filter system. The flow rate measuring device shall indicate gallons per minute (lpm) and shall be selected and installed to be accurate within plus or minus 10 percent of actual flow.

311.8 Instructions. Written operation and maintenance instructions shall be provided for the circulation system of public pools.

311.9 Hydrostatic pressure test. Circulation system piping, other than that integrally included in the manufacture of the pool or spa, shall be subjected to a hydrostatic pressure test of 25 pounds per square inch (psi) (172.4 kPa). This pressure shall be held for not less than 15 minutes.

SECTION 312
FILTERS

312.1 General. The provisions of this section apply to filters for pools and spas.

Exceptions:

1. Portable *residential* spas and portable *residential* exercise spas.
2. *Onground storable pools* supplied by the pool manufacturer as a kit that includes a filter that is in accordance with Section 704.

312.2 Design. Filters shall have a flow rating equal to or greater than the design flow rate of the system. Filters shall be installed in accordance with the manufacturer's instructions. Filters shall be designed so that filtration surfaces can be inspected and serviced.

312.3 Internal pressure. For pressure-type filters, a means shall be provided to allow the release of internal pressure.

312.3.1 Air release. Filters incorporating an automatic means of internal air release as the principal means of air release shall have one or more lids that provide a slow and safe release of pressure as a part of the design and shall have a manual air release in addition to an automatic release.

312.3.2 Separation tanks. A separation tank used in conjunction with a filter tank shall have a manual method of air release or a lid that provides for a slow and safe release of pressure as it is opened.

SECTION 313
PUMPS AND MOTORS

313.1 General. The provisions of this section apply to pumps and motors for pools and spas.

Exceptions:

1. Portable *residential* spas and portable *residential* exercise spas.
2. *Onground storable pools* supplied by the pool manufacturer as a kit that includes a pump and motor that is in accordance with Section 704.

TABLE 311.4.1
CIRCULATION SYSTEM FITTINGS

MATERIAL	STANDARD
Acrylonitrile butadiene styrene (ABS) plastic pipe	ASTM D1527
Chlorinated polyvinyl chloride (CPVC) plastic pipe and tubing	ASTM D2846; ASTM F437; ASTM F438; ASTM F439; CSA B137.6
Copper or copper-alloy tubing	ASME B16.15
Polyvinyl chloride (PVC) plastic pipe	ASTM D2464; ASTM D2466; ASTM D2467; CSA B137.2; CSA B137.3
Stainless steel pipe, Types 304, 304L, 316, 316L	ASTM A182; ASTM A403

313.2 Performance. A pump shall be provided for circulation of the pool water. The pump shall be capable of providing the flow required for filtering the pool water and filter cleaning, if applicable, against the total dynamic head developed by the complete system.

313.3 Intake protection. A cleanable strainer, skimmer basket, or screen shall be provided for pools and spas, upstream or as an integral part of circulation pumps, to remove solids, debris, hair, and lint on pressure filter systems.

313.4 Location. Pumps and motors shall be accessible for inspection and service in accordance with the manufacturer's specifications.

313.5 Safety. The design, construction, and installation of pumps and component parts shall be in accordance with the manufacturer's specifications.

313.6 Isolation valves. Shutoff valves shall be installed on the suction and discharge sides of pumps that are located below the waterline. Such valves shall be provided with access.

313.7 Emergency shutoff switch. An emergency shutoff switch shall be provided to disconnect power to recirculation and jet system pumps and air blowers. Emergency shutoff switches shall be: provided with access; located within sight of the pool or spa; and located not less than 5 feet (1524 mm) horizontally from the inside walls of the pool or spa.

Exception: *Onground storable pools*, permanent inground *residential* swimming pools, *residential* spas and *residential* water features.

313.8 Motor performance. Motors shall comply with UL 1004-1, UL 1081, CSA C22.2 No. 108 or the relevant motor requirements of UL 1563 or CSA C22.2 No. 218.1, as applicable.

SECTION 314 RETURN AND SUCTION FITTINGS

314.1 General. The provisions of this section apply to return and suction fittings for pools and spas

Exception: Portable *residential* spas and portable *residential* exercise spas.

314.2 Entrapment avoidance. Entrapment avoidance means shall be provided in accordance with Section 310.

314.3 Flow distribution. The suction outlet fitting assemblies, where installed, and the skimming systems shall each be designed to accommodate 100 percent of the circulation turnover rate.

314.3.1 Multiple systems. Where multiple systems are used in a single pool to meet this requirement, each subsystem shall proportionately be designed such that the maximum design flow rates cannot be exceeded during normal operation.

314.4 Return inlets. One return inlet shall be provided for every 300 square feet (27.9 m^2) of pool surface area, or fraction thereof.

Exception: *Onground storable pools.*

314.4.1 Design. Return and suction fittings for the circulation system shall be designed so as not to constitute a hazard to the bather.

314.5 Vacuum fittings. Where installed, *submerged vacuum fittings* shall be accessible and shall be located not greater than 12 inches (305 mm) below the water level.

SECTION 315 SKIMMERS

315.1 General. The provisions of this section apply to skimmers for pools and spas.

Exceptions:

1. Portable *residential* spas and portable *residential* exercise spas.
2. *Onground storable pools* supplied by the pool manufacturer as a kit that includes a skimming system that is in accordance with Section 704.

315.2 Required. A surface skimming system shall be provided for public pools and spas. Surface skimming systems shall be *listed* and *labeled* in accordance with NSF 50. Either a surface skimming system or perimeter overflow system shall be provided for permanent inground *residential* pools and permanent *residential* spas. Where installed, surface skimming systems shall be designed and constructed to create a skimming action on the pool water surface when the water level in the pool is within operational parameters.

Exceptions:

1. Class D public pools designed in accordance with Chapter 6.
2. Skimmers that are an integral part of a spa that has been *listed* and *labeled* in accordance with UL1563 shall not be required to be *listed* and *labeled* in accordance with NSF 50.

315.2.1 Circulation systems. Public pool circulation systems shall be designed to process not less than 100 percent of the turnover rate through skimmers.

315.3 Skimmer sizing. Where automatic surface skimmers are used as the sole overflow system, not less than one surface skimmer shall be provided for the square foot (square meter) areas, or fractions thereof, indicated in Table 315.3. Skimmers shall be located to maintain effective skimming action.

TABLE 315.3 SKIMMER SIZING TABLE

POOL OR SPA	AREA PER SKIMMER (SQ. FT)
Public pool	500
Residential pool	800
Spas (all types)	150

For SI: 1 square foot = 0.0929 m^2.

315.4 Perimeter coverage. Where a perimeter-type surface skimming system is used as the sole surface skimming system, the system shall extend around not less than 50 percent of the pool or spa perimeter.

315.4.1 Surge capacity. Where perimeter surface skimming systems are used, they shall be connected to a circulation system with a system surge capacity of not less than 1 gallon for each square foot (40.7 liters per square meter) of water surface. The capacity of the perimeter overflow system and related piping is permitted to be considered as a portion of the surge capacity.

315.5 Equalizers. Equalizers on skimmers shall be prohibited.

315.6 Hazard. Skimming devices shall be designed and installed so as not to create a hazard to the user.

SECTION 316 HEATERS

316.1 General. The provisions of this section apply to heaters for pools and spas.

Exception: Portable *residential* spas and portable *residential* exercise spas.

316.2 Listed and labeled. Heaters and hot water storage tanks shall be *listed* and *labeled* in accordance with the applicable standard listed in Table 316.2.

316.3 Sizing. Heaters shall be sized in accordance with the manufacturer's specifications.

316.4 Installation. Heaters shall be installed in accordance with the manufacturer's specifications and the *International Fuel Gas Code*, *International Mechanical Code*, *International Energy Conservation Code*, NFPA 70 or *International Residential Code*, as applicable in accordance with Section 102.7.1. Solar thermal water heaters shall be installed in accordance with Section 316.6.

316.4.1 Temperature. A means shall be provided to monitor water temperature.

316.4.2 Access prohibited. For public pools and spas, public access to controls shall be prohibited.

316.5 Heater circulation system. Heater circulation systems shall comply with Sections 316.5.1 and 316.5.2.

316.5.1 Water flow. Water flow through the heater bypass piping, back-siphonage protection, and the use of heat sinks shall be in accordance with the heater manufacturer's specifications.

316.5.2 Pump delay. Where required by the manufacturer, heaters shall be installed with an automatic device that will ensure that the pump continues to run after the heater shuts off for the time period specified by the manufacturer.

316.6 Solar thermal water heaters. Solar thermal heaters utilized for pools and spas shall comply with Sections 316.6.1 through 316.6.2.

316.6.1 Installation. Solar thermal water heaters shall be installed in accordance with the *International Mechanical Code* or *International Residential Code*, as applicable in accordance with Section 102.7.1.

316.6.2 Collectors and panels. Solar thermal collectors and panels shall be *listed* and *labeled* in accordance with ICC 901/SRCC 100 or ICC 900/SRCC 300. Collectors and panels shall be permanently marked with the manufacturer's name, model number, and serial number. Such markings shall be located on each collector in a position that is readily viewable after installation of the collector or panel.

SECTION 317 AIR BLOWER AND AIR INDUCTION SYSTEM

317.1 General. This section applies to devices and systems that induce or allow air to enter pools and spas either by means of a powered pump or passive design.

317.2 Backflow prevention. Air blower systems shall be equipped with backflow protection as specified in UL 1563 or CSA C22.2 No. 218.1.

317.3 Air intake source. Air intake sources shall not induce water, dirt or contaminants.

317.4 Sizing. Air induction systems shall be sized in accordance with the manufacturer's specifications.

317.5 Inspection and service. Air blowers shall be provided with access for inspection and service.

SECTION 318 WATER SUPPLY

318.1 Makeup water. Makeup water to maintain the water level and water used as a vehicle for sanitizers or other chemicals, for pump priming, or for other such additions, shall be from a potable water source.

318.2 Protection of potable water supply. Potable water supply systems shall be designed, installed and maintained so as to prevent contamination from nonpotable liquids, solids or gases being introduced into the potable water supply through cross-connections or other piping connections to the system. Means of protection against backflow in the potable

TABLE 316.2
WATER HEATERS

DEVICE	STANDARD
Electric water heater	UL 1261, UL 1563 or CSA C22.2 No. 218.1
Gas-fired water heater	ANSI Z21.56/CSA 4.7a
Heat exchanger	AHRI 400
Heat pump water heater	UL 1995, AHRI 1160, CSA C22.2 No. 236

water supply shall be provided through an air gap complying with ASME A112.1.2 or by a backflow prevention assembly in accordance with the *International Residential Code* or the *International Plumbing Code*, as applicable in accordance with Section 102.7.1.

318.3 Over-the-rim spouts. Over-the-rim spouts shall be located under a diving board, adjacent to a ladder, or otherwise shielded so as not to create a hazard. The open end of such spouts shall not have sharp edges and shall not protrude more than 2 inches (51 mm) beyond the edge of the pool. The open end shall be separated from the water by an air gap of not less than 1.5 pipe diameters measured from the pipe outlet to the rim.

SECTION 319 SANITIZING EQUIPMENT

319.1 Equipment standards. Sanitizing equipment installed in public pools and spas shall be capable of introducing the quantity of sanitizer necessary to maintain the appropriate levels under all conditions of intended use.

319.2 Chemical feeders. Where installed, chemical feed systems shall be installed in accordance with the manufacturer's specifications. Chemical feed pumps shall be wired so that they cannot operate unless there is adequate return flow to disburse the chemical throughout the pool or spa as designed.

SECTION 320 WASTEWATER DISPOSAL

320.1 Backwash water or draining water. Backwash water and draining water shall be discharged to the sanitary or storm sewer, or into an *approved* disposal system on the premise, or shall be disposed of by other means *approved* by the state or local authority. Direct connections shall not be made between the end of the backwash line and the disposal system. Drains shall discharge through an air gap.

320.2 Water salvage. Filter backwash water shall not be returned to the vessel except where the backwash water has been filtered to remove particulates, treated to eliminate coli form bacteria and waterborne pathogens, and such return has been *approved* by the state or local authority.

320.3 Waste post treatment. Where necessary, filter backwash water and drainage water shall be treated chemically or through the use of settling tanks to eliminate or neutralize chemicals, diatomaceous earth, and contaminants in the water that exceed the limits set by the state or local effluent discharge requirements.

SECTION 321 LIGHTING

321.1 General. The provisions of Sections 321.2 and 321.3 shall apply to lighting for public pools and spas. The provisions of Section 321.4 shall apply to lighting for *residential* pools and spas.

321.2 Artificial lighting required. When a pool is open during periods of low natural illumination, artificial lighting shall be provided so that all areas of the pool, including all suction outlets on the bottom of the pool, will be visible. Illumination shall be sufficient to enable a lifeguard or other persons standing on the deck or sitting on a lifeguard stand adjacent to the pool edge to determine if a pool user is lying on the bottom of the pool and that the pool water is transparent and free from cloudiness.

These two conditions shall be met when all suction outlets are visible from the edge of the deck at all times when artificial lighting is illuminated and when an 8-inch-diameter (152 mm) black disk, placed at the bottom of the pool in the deepest point, is visible from the edge of the pool deck at all times when artificial lighting is illuminated.

321.2.1 Pool and deck illumination. Overhead lighting, underwater lighting or both shall be provided to illuminate the pool and adjacent deck areas. The lighting shall be *listed* and *labeled*. The lighting shall be installed in accordance with NFPA 70.

321.2.2 Illumination intensity. For outdoor pools, any combination of overhead and underwater lighting shall provide *maintained illumination* not less than 10 horizontal foot-candles (10 lumens per square foot) [108 lux] at the pool water surface. For indoor pools, any combination of overhead and underwater lighting shall provide *maintained illumination* of not less than 30 horizontal foot-candles (30 lumens per square foot) [323 lux] at the pool water surface. Deck area lighting for both indoor and outdoor pools shall provide *maintained illumination* of not less than 10 horizontal foot-candles (10 lumens per square foot) [108 lux] at the walking surface of the deck.

321.2.3 Underwater lighting. Underwater lighting shall provide not less than 8 horizontal foot-candles (8 lumens per square foot) [86 lux] at the pool water surface area, or not less than a total wattage of $^1/_2$ watt/ft^2 (5.4 watts/m^2) of pool water surface for incandescent underwater lighting where the fixtures and lamps are rated in watts.

Exception: The requirement of this section shall not apply where overhead lighting provides not less than 15 foot-candles (15 lumens per square foot) [161 lux] of *maintained illumination* at the pool water surface, the overhead lighting provides visibility, without glare, of all areas of the pool, and the requirements of Section 321.2.2 are met or exceeded.

321.3 Emergency illumination. Public pools and public pool areas that operate during periods of low illumination shall be provided with emergency lighting that will automatically turn on to permit evacuation of the pool and securing of the area in the event of power failure. Emergency lighting facilities shall be arranged to provide initial illumination that is not less than 0.1 foot-candle (0.1 lumen per square foot) [1 lux] measured at any point on the water surface and at any point on the walking surface of the deck, and not less than an average of 1 foot-candle (1 lumen per square foot) [11 lux]. At the end of the emergency lighting time duration, the illumination level shall be not less than 0.06 foot-candle (0.06 lumen per square foot) [0.65 lux] measured at any point on the water surface and at any point on the walking surface of the deck, and not less than an average of 0.6 foot-candle (0.6 lumen per square foot) [6.46 lux]. A maximum-to-minimum illumination uniformity ratio of 40 to 1 shall not be exceeded.

321.4 Residential pool and deck illumination. Where lighting is installed for, and in, *residential* pools and permanent *residential* spas, such lighting shall be installed in accordance with NFPA 70 or the *International Residential Code*, as applicable in accordance with Section 102.7.1.

SECTION 322
LADDERS AND RECESSED TREADS

322.1 General. Ladders and recessed treads shall comply with the provisions of this section and the applicable provisions of Chapters 4 through 10 based on the type of pool or spa.

322.2 Outside diving envelope. Where installed, steps and ladders shall be located outside of the minimum diving water envelope as indicated in Figure 322.2.

322.3 Ladders. Ladder treads shall have a uniform horizontal depth of not less than 2 inches (51 mm). There shall be a uniform distance between ladder treads, with a distance of not less than 7 inches (178 mm) and not greater than 12 inches (305 mm). The top tread of a ladder shall be located not greater than 12 inches (305 mm) below the top of the deck or coping. Ladder treads shall have slip-resistant surfaces.

322.3.1 Wall clearance. There shall be a clearance of not less than 3 inches (76 mm) and not greater than 6 inches (152 mm) between the pool wall and the ladder.

322.3.2 Handrails and handholds. Ladders shall be provided with two handholds or two handrails. The clear distance between ladder handrails shall be not less than 17 inches (432 mm) and not greater than 24 inches (610 mm).

322.4 Recessed treads. Recessed treads shall have a minimum depth of not less than 5 inches (127 mm) and a width of not less than 12 inches (305 mm). The vertical distance between the pool coping edge, deck, or step surface and the uppermost recessed tread shall be not greater than 12 inches (305 mm). Recessed treads shall have slip-resistant surfaces.

322.4.1 Vertical spacing. Recessed treads at the centerline shall have a uniform vertical spacing of not less than 7 inches (178 mm) and not greater than 12 inches (305 mm).

322.4.2 Drainage. Recessed treads shall drain into the pool.

322.4.3 Handrails and grab rails. Recessed treads shall be provided with a handrail or grab rail on each side of the treads. The clear distance between handrails and grab rails shall be not less than 17 inches (432 mm) and not greater than 24 inches (610 mm).

SECTION 323
SAFETY

323.1 Handholds required. Where the depth below the *design waterline* of a pool or spa exceeds 42 inches (1067 mm), handholds along the perimeter shall be provided. Handholds shall be located at the top of deck or coping.

Exceptions:

1. Handholds shall not be required where an underwater bench, seat or swimout is installed.
2. Handholds shall not be required for wave action pools and action rivers.

323.1.1 Height above water. Handholds shall be located not more than 12 inches (305 mm) above the *design waterline.*

323.1.2 Handhold type. Handholds shall be one or more of the following:

1. Top of pool deck or coping.
2. Secured rope.
3. Rail.
4. Rock.
5. Ledge.
6. Ladder.
7. Stair step.
8. Any design that allows holding on with one hand while at the side of the pool.

323.1.3 Handhold spacing. Handholds shall be horizontally spaced not greater than 4 feet (1219 mm) apart.

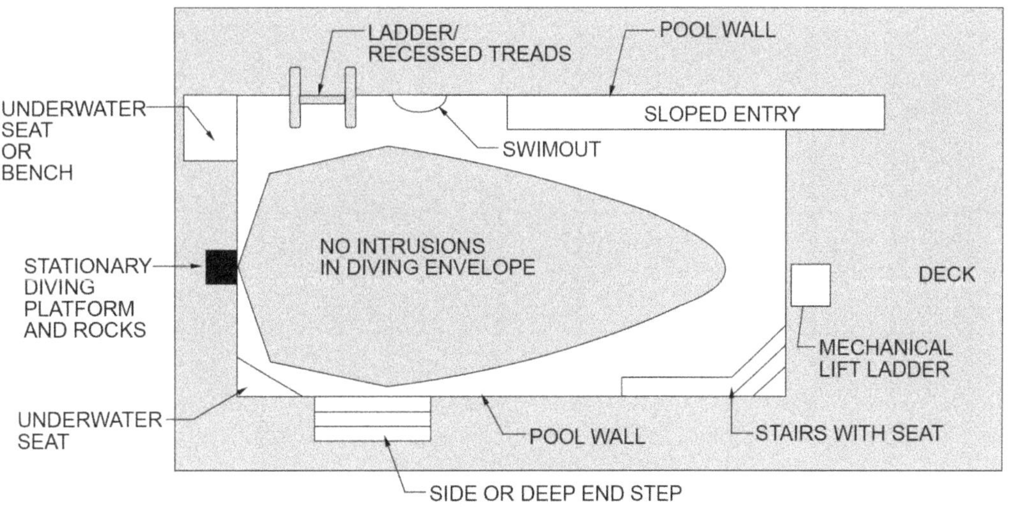

FIGURE 322.2
MINIMUM WATER DIVING ENVELOPE

323.2 Handrails. Where handrails are installed, they shall conform to this section.

323.2.1 Height. The top of the gripping surface of handrails for public pools and public spas shall be 34 inches (864 mm) to 38 inches (965 mm) above the ramp or step surface as measured at the nosing of the step or finished surface of the slope. The top of the gripping surface of handrails for *residential* pools and *residential* spas shall be 30 inches (762 mm) to 38 inches (965 mm) above the ramp or step surface as measured at the nosing of the step or finished surface of the slope.

323.2.2 Material. Handrails shall be made of corrosion-resistant materials.

323.2.3 Nonremovable. Handrails shall be installed so that they cannot be removed without the use of tools.

323.2.4 Leading edge distance. The leading edge of handrails for stairs, pool entries and exits shall be located not greater than 18 inches (457 mm) from the vertical face of the bottom riser.

323.2.5 Diameter. The outside diameter or width of *handrails* shall be not less than $1^1/_4$ inches (32 mm) and not greater than 2 inches (51 mm).

323.3 Obstructions and entrapment avoidance. There shall not be obstructions that can cause the user to be entrapped or injured. Types of entrapment include, but are not limited to, wedge or pinch-type openings and rigid, nongiving cantilevered protrusions.

CHAPTER 4

PUBLIC SWIMMING POOLS

User note:

About this chapter: *Chapter 4 has regulations for public swimming pools. Where diving boards are present, the chapter provides information regarding the minimum diving water dimensions. Requirements for exiting pools, decks, circulation systems and depth markers are provided. Special features of pools such as rest ledges, swimouts and underwater seats and benches are regulated by this chapter.*

SECTION 401 GENERAL

401.1 Scope. The provisions of this chapter shall apply only to Class A, Class B, Class C, Class E and Class F public swimming pools.

401.2 Intent. The provisions in this chapter shall govern the design, equipment, operation, warning signs, installation, sanitation, new construction, and alteration specific to the types of public swimming pools indicated in Section 401.1.

401.3 Chapter 3 compliance required. In addition to the requirements of this chapter, public swimming pools shall comply with the requirements of Chapter 3.

401.4 Dimensional tolerances. Finished pool dimensions, for other than Class A pools, shall be held within the construction tolerances shown in Table 401.4. Other dimensions, unless otherwise specified, shall have a tolerance of ± 2 inches (51 mm).

401.4.1 Class A pool tolerances. Dimensional tolerances for Class A pools shall be determined by the authority that provides the accreditation of the pool for competitive events.

401.5 Floor slope. Except where required to meet the accessibility requirements in accordance with Section 307.9, the slope of the floor in the shallow area of a pool shall not exceed 1 unit vertical in 10 units horizontal (10-percent slope) for Class C pools and 1 unit vertical in 12 units horizontal (8-percent slope) for Class B pools. The slope limit shall apply in any direction to the point of the first slope change, where a slope change exists. The point of the first slope change shall be defined as the point at which the floor slope exceeds 1 unit vertical in 10 units horizontal (10-percent slope) for Class C pools and 1 unit vertical in 12 units horizontal (8-percent slope) for Class B pools.

401.6 Dimensions for Class A pools. Class A pools shall be designed and constructed with the dimensions determined by the authority that provides the accreditation of the pool for competitive events.

SECTION 402 DIVING

402.1 General. This section covers diving requirements for Class B, Class C, and Class E pools. Manufactured and fabricated diving equipment and appurtenances shall not be installed on Type O pools.

402.2 Manufactured and fabricated diving equipment. Manufactured and fabricated diving equipment shall be in accordance with this section and shall be designed for swimming pool use.

402.3 Installation. The installation of manufactured diving equipment shall be in accordance with Sections 402.3 through 402.14. Manufactured diving equipment shall be located in the deep area of the pool so as to provide the minimum dimensions shown in Table 402.12 and shall be installed in accordance with the manufacturer's instructions. Installation and use instructions for manufactured diving equipment shall be provided by the manufacturer and shall

**TABLE 401.4
CONSTRUCTION TOLERANCES**

DESIGN ASPECT	CONSTRUCTION TOLERANCE
Depth–deep area, including diving area	± 3 inches
Depth–shallow area	± 2 inches
Length–overall	± 3 inches
Step treads & risers	± $^1/_2$ inch
Wall slopes	± 3 degrees
Waterline–pools with adjustable weir skimmers	± $^1/_4$ inch
Waterline–pools with nonadjustable skimming systems (gutters)	± $^1/_8$ inch
Width–overall	± 3 inches
All dimensions not otherwise specified herein	± 2 inches

For SI: 1 inch = 25.4 mm, 1 degree = 0.017 radians.

specify the minimum diving water envelope dimensions required for each diving board and diving stand combination. The manufacturer's instructions shall refer to the water envelope type by dimensionally relating their products to Point A on the diving water envelopes shown in Table 402.12. The diving board manufacturer shall specify which boards fit on the design pool geometry types as indicated in Table 402.12.

402.4 Slip resistance. Diving equipment shall have slip-resistant walking surfaces.

402.5 Point A. For the application of Table 402.12, Point A shall be the point from which dimensions of width, length and depth are established for the minimum diving water envelope. If the tip of the diving board or diving platform is located at a distance of WA (see Figure 804.1) or greater from the deep end wall and the water depth at that location is equal to or greater than the water depth requirement at Point A, the point on the water surface directly below the center of the tip of the diving board or diving platform shall be identified as Point A.

402.6 Location of pool features in a diving pool. Where a pool is designed for use with diving equipment, the location of steps, pool stairs, ladders, underwater benches, underwater ledges, special features and other accessory items shall be outside of the minimum diving water envelope. See Figure 322.2.

402.7 Stationary diving platforms and diving rocks. Where stationary diving platforms and diving rocks are built on site, flush with the wall and located in the diving area of the pool, Point A shall be in front of the wall at the platform or diving rock centerline.

402.8 Location of diving equipment. Manufactured and fabricated diving equipment shall be located so that the tip of the board or platform is located directly above Point A as defined by Section 402.5.

402.9 Elevation. The maximum elevation of a diving board above the *design waterline* shall be in accordance with the manufacturer's instructions.

402.10 Platform height above waterline. The height of an *approved* stationary diving apparatus, platform, or diving rock above the *design waterline* shall not exceed the limits of the manufacturer's specifications or the limits of the design prepared by a design professional.

402.11 Clearance. The diving equipment manufacturer shall specify the minimum headroom required above the tip of the board.

402.12 Water envelopes. The minimum diving water envelopes shall be in accordance with Table 402.12.

402.13 Ladders for diving equipment. Ladders shall be provided with two grab rails or two handrails. There shall be a uniform distance between ladder treads, with a 7-inch (178 mm) minimum distance and a 12-inch (305 mm) maximum distance.

Exception: The distance between treads for the top and bottom riser can vary but shall be not less than 7 inches (178 mm) and not greater than 12 inches (305 mm).

402.14 Springboard fall protection guards. Springboards located at a height greater than 5 feet (1524 mm) above the pool deck shall have a fall protection guard on each side of

TABLE 402.12
MINUMUM DIVING WATER ENVELOPES
(SEE FIGURE 402.12)

POOL TYPE	MINIMUM DIMENSIONS								MINIMUM WIDTH OF POOL AT:		
	D_1	D_2	R	L_1	L_2	L_3	L_4	L_5	Pt. A	Pt. B	Pt. C
VI	7'-0"	8'-6"	5'-6"	2'-6"	8'-0"	10'-6"	7'-0"	28'-0"	16'-0"	18'-0"	18'-0"
VII	7'-6"	9'-0"	6'-0"	3'-0"	9'-0"	12'-0"	4'-0"	28'-0"	18'-0"	20'-0"	20'-0"
VIII	8'-6"	10'-0"	7'-0"	4'-0"	10'-0"	15'-0"	2'-0"	31'-0"	20'-0"	22'-0"	22'-0"
IX	11'-0"	12'-0"	8'-6"	6'-0"	10'-6"	21'-0"	0	37'-6"	22'-0"	24'-0"	24'-0"

For SI: 1 inch = 25.4 mm, 1 foot = 304.8 mm.

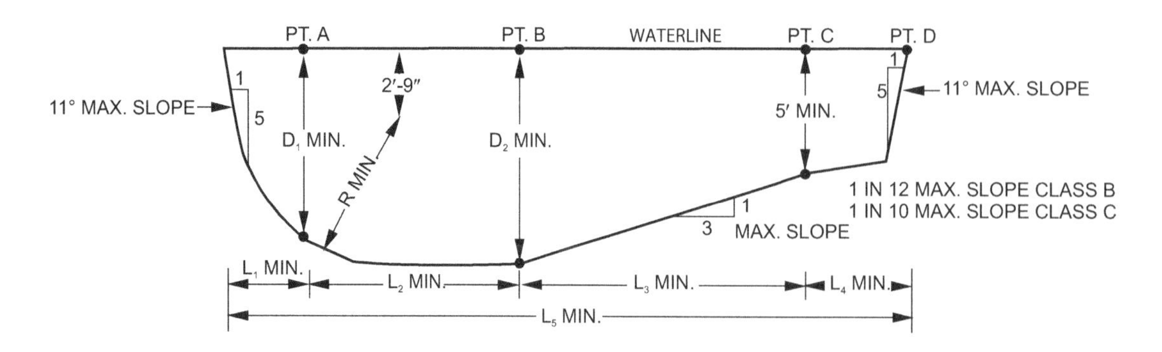

For SI: 1 degree = 0.017 rad, 1 inch = 25.4 mm, 1 foot = 304.8 mm.

FIGURE 402.12
(MINIMUM DIVING WATER ENVELOPES)
CONSTRUCTION DIMENSIONS FOR WATER ENVELOPES FOR CLASS B AND CLASS C POOLS

TABLE 403.1
MAXIMUM BATHER LOAD

POOL/DECK AREA	SHALLOW INSTRUCTIONAL OR WADING AREAS	DEEP AREA (NOT INCLUDING THE DIVING AREA)	DIVING AREA (PER EACH DIVING BOARD)
Pools with minimum deck area	15 sq. ft. per user	20 sq. ft. per user	300 sq. ft.
Pools with deck area at least equal to water surface area	12 sq. ft. per user	15 sq. ft. per user	300 sq. ft.
Pools with deck area at least twice the water surface area	8 sq. ft. per user	10 sq. ft. per user	300 sq. ft.

For SI: 1 square foot = 0.09 square meters.

the springboard. The design and the selection of the materials of construction of the fall protection guards shall be determined by the manufacturer of the springboard support structure. The installation and maintenance of the fall protection guards shall be in accordance with the fall protection guard manufacturer's instructions.

SECTION 403 BATHER LOAD

403.1 Maximum bather load. The maximum bather load of Class B and Class C pools shall be in accordance with Table 403.1.

SECTION 404 REST LEDGES

404.1 Rest ledges. Rest ledges along the pool walls are permitted. They shall be not less than 4 feet (1220 mm) below the water surface. Where a ledge is provided, the width of the ledge shall be not less than 4 inches (102 mm) and not greater than 6 inches (152 mm).

SECTION 405 WADING POOLS

405.1 Wading pools. Class F wading pools shall be separate pools with an independent circulation system, shall be physically separated from the main pool and shall be constructed in accordance with Sections 405.2 through 405.6.

405.2 Nonentry areas. The areas where the water depth at the edge of the pool exceeds 9 inches (229 mm) shall be considered to be nonentry areas.

405.3 Floor slope. The floors of wading pools shall be uniform and sloped with a maximum slope of 1 unit vertical in 12 units horizontal (8-percent slope).

405.4 Maximum depth. The water depth shall not exceed 18 inches (457 mm).

405.5 Distance from deck to waterline. The maximum distance from the top of the deck to the waterline shall not exceed 6 inches (152 mm).

405.6 Suction entrapment avoidance. Wading pools shall not have suction outlets. Skimmers or overflow *gutters* shall be installed and shall accommodate 100 percent of the circulation system flow rate.

SECTION 406 DECKS AND DECK EQUIPMENT

406.1 General. Decks shall comply with the provisions of Section 306, except as otherwise required in this section.

406.2 Pool perimeter access. A deck or unobstructed access shall be provided for not less than 90 percent of the pool perimeter.

406.3 Deck clearance. Decking not less than 4 feet (1219 mm) in width shall be provided on the sides and rear of any diving equipment. A deck clearance of 4 feet (1219 mm) shall be provided around all other deck equipment.

406.4 Decks between pools and spas. Decks between pools, spas or any combination of pools and spas, shall have a width of not less than 6 feet (1829 mm).

406.5 Deck covering. Walking surfaces of decks within 4 feet (1219 mm) of a pool or spa shall be slip resistant.

406.6 Distances above diving boards. A completely unobstructed minimum distance above the tip of the diving board shall be specified by the diving equipment manufacturer.

406.7 Dimensional requirements. Public pools with diving equipment of 39 inches (991 mm) or greater in height, and pools designed for springboard or platform diving, shall comply with the dimensional design requirements of the diving equipment manufacturer or the authority that governs such pools.

406.8 Diving equipment. Diving equipment shall be installed in accordance with the manufacturer's specifications.

406.8.1 Label. A label shall be permanently affixed to the diving equipment or jump board in a readily visible location and shall include all of the following:

1. The minimum diving water envelope required for each diving board and diving stand combination.
2. Manufacturer's name and address.
3. Manufacturer's identification and date of manufacture.
4. The maximum allowable weight of the user.

406.8.2 Use instructions. The diving equipment manufacturer shall provide diving equipment use instructions.

406.8.3 Tread surface. Diving equipment shall have slip-resistant tread surfaces.

406.8.4 Supports for diving equipment. Supports, platforms, stairs, and ladders for diving equipment shall be designed to carry the anticipated loads. Stairs and ladders shall be of corrosion-resistant materials, shall be easily cleanable and shall have slip-resistant treads. Diving stands higher than 21 inches (533 mm), measured from the deck to the top back end of the board, shall be provided with stairs or a ladder. Step treads shall be self-draining.

406.8.5 Guardrails. Diving equipment 39 inches (991 mm) or greater in height shall be provided with a top guardrail. Such guardrail shall extend not less than 30 inches (762 mm) above the diving board and extend to the edge of the pool wall.

406.9 Starting blocks. In new construction or substantial alteration, starting blocks intended for competitive swimming shall be located at a water depth of not less than 5 feet (1524 mm).

406.10 Swimming pool slides. Swimming pool slides shall comply with the requirements of 16 CFR, Part 1207. The manufacturer of the slide shall provide installation and use instructions for the slide. Slides shall be installed in accordance with the manufacturer's instructions.

406.11 Play and water activity equipment. Play and water activity equipment shall be installed in accordance with the manufacturer's instructions.

SECTION 407 CIRCULATION SYSTEMS

407.1 General. Circulation systems for pools shall comply with Section 311 and the provisions of this section.

407.2 Turnover. Circulation equipment shall be sized to turn over the entire water capacity of the pool as specified in Table 407.2. The system shall be designed to provide the required turnover rate based on the maximum pressure and flow rate recommended by the manufacturer of the filter with clean filter media.

SECTION 408 FILTERS

408.1 General. Filters shall be designed in accordance with Section 312, except as otherwise required in this section.

408.2 Air release warning. The following statement shall be posted in a conspicuous location within the areas of the air release:

> DO NOT START THE SYSTEM AFTER MAINTENANCE WITHOUT FIRST PROPERLY REASSEMBLING THE FILTER AND SEPARATION TANK AND OPENING ALL AIR RELEASE VALVES.

SECTION 409 SPECIFIC SAFETY FEATURES

409.1 Handholds. Handholds shall comply with the provisions of Section 323.

409.2 Depth markers. Depth markers shall be provided in accordance with Sections 409.2.1 through 409.2.8.

409.2.1 Where required. Depth markers shall be installed at the maximum and minimum water depths and at all points of slope change. Depth markers shall be installed at water depth increments not to exceed 2 feet (607 mm). Depth markers shall be spaced at intervals not to exceed 25 feet (7620 mm).

409.2.2 Marking of depth. The depth of water in feet (meters) shall be plainly and conspicuously marked on the vertical pool wall at or above the waterline.

Exception: Pools with a vanishing edge and rim flow *gutters*.

409.2.3 Depth accuracy. Depth markers shall indicate the actual pool depth within ± 3 inches (76 mm), at normal operating water level where measured 3 feet (914 mm) from the pool wall or at the tangent point where the cove radius meets the floor, whichever is deeper.

409.2.4 Position on pool wall. Depth markers on the vertical pool wall shall be positioned to be read from the waterside. Depth markers shall be placed so as to allow as much of the numbers to be visible above the waterline as possible.

409.2.5 Position on deck. Depth markers on the deck shall be located within 18 inches (457 mm) of the water edge and positioned to be read while standing on the deck facing the water.

409.2.6 Horizontal markers. Horizontal depth markers shall be slip resistant.

409.2.7 Uniform distribution. Depth markers shall be distributed uniformly on both sides and both ends of the pool.

409.2.8 Numbers and letters. Depth markers shall be not less than 4 inches (102 mm) in height. The color of the numbers shall contrast with the background on which they are applied and the color shall be of a permanent nature. The lettering shall spell out the words "feet" and "inches" or abbreviate them as "Ft." and "In." respectively. Where displayed in meters in addition to feet and inches, the word meter shall be spelled out or abbreviated as "M."

409.3 No diving symbol. Where the pool depth is 5 feet (1524 mm) or less, the "No Diving" symbol shall be displayed. The symbol shall be placed on the deck at intervals of not greater than 25 feet (7620 mm) and directly adjacent to a depth marker. Additional signage shall be in accordance with NEMA Z535.

TABLE 407.2 TURNOVER RATE

SWIMMING POOL CATEGORY	TURNOVER RATE IN HOURS
Class A, B, and C pools	Hours equal $1^1/_2$ times the average depth of pool in feet not to exceed 6 hours
Wading pools	1

For SI: 1 foot = 304.8 mm.

409.4 Lifesaving equipment. Public pool Classes A, B, and C shall be provided with lifesaving equipment in accordance with Sections 409.4.1 through 409.4.3. Such lifesaving equipment shall be visually conspicuous and conveniently located at all times.

409.4.1 Accessory pole. A swimming pool accessory pole not less than 12 feet (3658 mm) in length and including a body hook shall be provided.

409.4.2 Throwing rope. A throwing rope attached to a ring buoy or similar flotation device shall be provided. The rope shall be not less than $^1/_4$ inch (6.4 mm) in diameter and shall have a length of not less than $1^1/_2$ times the maximum width of the pool or 50 feet (15 240 mm), whichever is less. A ring buoy shall have an outside diameter of not less than 15 inches (381 mm).

409.4.3 Emergency response units. Pools covered by this chapter shall be provided with first aid equipment, including a first aid kit. First aid equipment and kits shall be located in an accessible location.

SECTION 410 SANITARY FACILITIES

410.1 Toilet facilities. Class A and B pools shall be provided with toilet facilities having the required number of plumbing fixtures in accordance with the *International Building Code* or the *International Plumbing Code*.

SECTION 411 SPECIAL FEATURES

411.1 Entry and exit. Pools shall have not less than two means of entry and exit that are located so as to serve both ends of a pool. Pool lifts, transfer walls and transfer systems that provide for pool entry and exit by persons with physical disabilities in accordance with Section 307.1.4 shall not be counted as the means of entry or exit that is required by this section.

411.1.1 Natural entry. Where areas have water depths of 24 inches (607 mm) or less at the pool wall, such areas shall be considered to be providing their own natural mode for entry and exit.

Exception: Wading pools as outlined in Section 405.

411.1.2 Shallow area. A means of entry and exit shall be provided in shallow areas of pools and shall consist of pool stairs, a ramp or a beach entry.

411.1.3 Deep area. The means of entry and exit in the deep area of pools shall consist of one of the following:

1. Steps/stairs.
2. Ladders.
3. Grab rails with recessed treads.
4. Ramps.
5. Beach entries.
6. Swimouts.
7. Other designs that provide the minimum utility as specified in this code.

411.1.4 Pools greater than 30 feet wide. Swimming pools greater than 30 feet (9144 mm) in width shall be provided with entries and exits on each side of the deep area of the pool. The entries and exits on the sides of the deep area of a pool shall be located not more than 82 feet (25 m) apart.

411.1.5 Diving envelope. Where the pool is designed for use with diving equipment, the entries and exits, pool stairs, ladders, underwater benches, special features and other accessories shall be located outside of the minimum diving water envelope indicated in Figure 322.2.

411.1.6 Treads. Treads shall have slip-resistant surfaces.

411.2 Pool stairs. The design and construction of stairs extending into the pool in either shallow or deep water, including recessed pool stairs, shall comply with Sections 411.2.1 through 411.2.4.

411.2.1 Tread dimensions and area. Treads shall be not less than 24 inches (607 mm) at the leading edge. Treads shall have an unobstructed surface area of not less than 240 square inches (0.154 m^2) and an unobstructed horizontal depth of not less than 10 inches (254 mm) at the centerline.

411.2.2 Risers. Risers, except for the bottom riser, shall have a uniform height of not greater than 12 inches (305 mm) measured at the centerline. The bottom riser height is allowed to vary to the floor.

411.2.3 Top tread. The vertical distance from the pool coping, deck, or step surface to the uppermost tread shall be not greater than 12 inches (305 mm).

411.2.4 Bottom tread. Where stairs are located in water depths greater than 48 inches (1219 mm), the lowest tread shall be not less than 48 inches (1219 mm) below the deck and shall be recessed in the pool wall.

411.3 Shallow end detail for beach and sloping entries. Sloping entries used as a pool entrance shall have a maximum slope of 1 unit vertical in 10 units horizontal (10-percent slope).

411.3.1 Benches and steps. Where benches are used in conjunction with sloping entries, the vertical riser distance shall not exceed 12 inches (305 mm). Where steps are used in conjunction with sloping entries, the requirements of Section 411.2 shall apply.

411.3.2 Vertical drops. A vertical drop exceeding 12 inches (305 mm) within a sloping entry shall be provided with a handrail.

411.3.3 Surfaces. Beach and sloping entry surfaces shall be of slip-resistant materials.

411.4 Pool ladder design and construction. The design and construction of ladders shall comply with Section 322.

411.5 Underwater seats, benches, and swimouts. The design and construction of underwater seats, benches, and *swimouts* shall comply with Sections 411.5.1 and 411.5.2.

411.5.1 Swimouts. Swimouts, located in either the deep or shallow area of a pool, shall comply with all of the following:

1. The horizontal surface shall be not greater than 20 inches (508 mm) below the waterline.

2. An unobstructed surface shall be provided that is equal to or greater than that required for the top tread of the pool stairs in accordance with Section 411.2.
3. Where used as an entry and exit access, swimouts shall be provided with steps that comply with the pool stair requirements of Section 411.2.
4. The leading edge shall be visibly set apart.

411.5.2 Underwater seats and benches. Underwater seats and benches, whether used alone or in conjunction with pool stairs, shall comply with all of the following:

1. The horizontal surface shall be not greater than 20 inches (508 mm) below the waterline.
2. An unobstructed surface shall be provided that is not less than 10 inches (254 mm) in depth and not less than 24 inches (607 mm) in width.
3. Underwater seats and benches shall not be used as the required entry and exit access.
4. Where underwater seats are located in the deep area of the pool where manufactured or constructed diving equipment is installed, such seats shall be located outside of the minimum diving water envelope for diving equipment.
5. The leading edge shall be visually set apart.
6. The horizontal surface shall be at or below the waterline.
7. A tanning ledge or sun shelf used as the required entry and exit access shall be located not greater than 12 inches (305 mm) below the waterline.

SECTION 412
SIGNAGE

412.1 Safety signage. Safety signage advising on the danger of diving into *shallow areas* and on the prevention of drowning shall be provided as required by the authority that governs such pools. Safety signage shall be as shown in Figure 412.1 or similar thereto.

412.2 Emergency telephone signs. A sign indicating the location of the nearest landline telephone that can be used to call emergency services shall be posted within sight of the main entry into a pool facility. The sign shall indicate the telephone numbers, including area code, that can be called for emergency services including, but not limited to, police, fire, ambulance and rescue services. If "9-1-1" telephone service is available for any of those services, "9-1-1" shall be indicated next to the telephone number for such services. The sign shall include the street address and city where the pool is located. The nearest landline telephone indicated by the sign shall be one that can be used free of charge to call for emergency services. A sign with the telephone number and address information required by this section shall be posted within sight of the landline telephone.

412.3 Sign placement. Signs shall be positioned for effective visual observation by users as required by the authority that governs such pools.

412.4 Emergency shutoff switch. Signs shall be posted that clearly indicate the location of the pump emergency shutoff switch. Such switch shall be clearly identified as the pump emergency shutoff switch.

Actual Size:
11″ x 13- 3/8″
16″ x 18-1/2″
18-3/8″ x 24″

For SI: 1 inch = 25.4 mm.

FIGURE 412.1
SAFETY SIGN

CHAPTER 5

PUBLIC SPAS AND PUBLIC EXERCISE SPAS

User note:

About this chapter: *Chapter 5 regulates the depth, seating depth and floor slope of public pools and public exercise spas. Suction fitting, heater and depth marker requirements are also included in this chapter.*

SECTION 501 GENERAL

501.1 Scope. This chapter shall govern the design, installation, construction and repair of public spas and exercise spas regardless of whether a fee is charged for use.

501.2 General. In addition to the requirements of this chapter, public spas and public exercise spas shall comply with the requirements of Chapter 3.

SECTION 502 MATERIALS

502.1 Pumps and motors. Pumps and motors shall be *listed* and *labeled* for use in spas.

SECTION 503 STRUCTURE AND DESIGN

503.1 Water depth. The maximum water depth for spas shall be 4 feet (1219 mm) measured from the *design waterline* except for spas that are designed for special purposes and *approved* by the authority having jurisdiction. The water depth for exercise spas shall not exceed 6 feet 6 inches (1981 mm) measured from the *design waterline*.

503.2 Multilevel seating. Where multilevel seating is provided, the maximum water depth of any seat or sitting bench shall be 28 inches (711 mm) measured from the *design waterline* to the lowest measurable point.

503.3 Floor slope. The slope of the floor shall not exceed 1 unit vertical in 12 units horizontal (8.3-percent slope). Where multilevel floors are provided, the change in depth shall be indicated.

SECTION 504 PUMPS AND MOTORS

504.1 Emergency shutoff switch. One emergency shutoff switch shall be provided to disconnect power to circulation and jet system pumps and air blowers. Emergency shutoff switches shall be accessible, located within sight of the spa and shall be located not less than 5 feet (1524 mm) but not greater than 10 feet (3048 mm) horizontally from the inside walls of the spa.

504.1.1 Alarms. Emergency shutoff switches shall be provided with an audible alarm rated at not less than 80 decibel sound pressure level and a light near the spa that will operate continuously until deactivated when the shutoff switch is operated. The following statements shall appear on a sign that is posted in a location that is visible from the spa:

ALARM INDICATES SPA PUMPS OFF. DO NOT USE SPA WHEN ALARM SOUNDS AND LIGHT IS ILLUMINATED UNTIL ADVISED OTHERWISE.

SECTION 505 RETURN AND SUCTION FITTINGS

505.1 Return fittings. Return fittings shall be provided and arranged to facilitate a uniform circulation of water and maintain a uniform sanitizer residual throughout the entire spa or exercise spa.

505.2 Suction fittings. Suction fittings shall be in accordance with Sections 505.2.1 through 505.2.4.

505.2.1 Testing and certification. Suction fittings shall be *listed* and *labeled* in accordance with APSP 16.

505.2.2 Installation. Suction fittings shall be sized and installed in accordance with the manufacturer's specifications. Spas and exercise spas shall not be used or operated if the suction outlet cover is missing, damaged, broken or loose.

505.2.3 Outlets per pump. Suction fittings shall be provided in accordance with Section 310.

505.2.4 Submerged vacuum fittings. *Submerged vacuum fittings* shall be in accordance with Section 310.

SECTION 506 HEATER AND TEMPERATURE REQUIREMENTS

506.1 General. This section pertains to fuel-fired and electric appliances used for heating spa or exercise spa water.

506.2 Water temperature controls. Components provided for water temperature controls shall be suitable for the intended application.

506.2.1 Water temperature regulating controls. Water temperature regulating controls shall comply with UL 873 or UL 372. A means shall be provided to indicate the water temperature in the spa.

Exception: Water temperature regulating controls that are integral to the heating appliance and *listed* in accordance with the applicable end use appliance standard.

506.2.2 Water temperature limiting controls. Water temperature limiting controls shall comply with UL 873 or UL 372. Water temperature at the heater return outlet shall not exceed 140°F (60°C).

SECTION 507 WATER SUPPLY

507.1 Water temperature. The temperature of the incoming makeup water shall not exceed 104°F (40°C).

SECTION 508 SANITIZING, OXIDATION EQUIPMENT AND CHEMICAL FEEDERS

508.1 Automatic controllers. Where an automatic controller is installed on a spa or hot tub for public use, the controller shall be installed with an automatic pH and an oxidation reduction potential controller *listed* and *labeled* in compliance with NSF 50.

SECTION 509 SAFETY FEATURES

509.1 Instructions and safety signs. Instructions and safety signage shall comply with the requirements of the local jurisdiction. In the absence of local requirements, safety signs and instructions shall comply with UL 1563 or CSA C22.2 No. 218.1.

509.2 Operational signs. Operational signs shall include, but not be limited to, the following messages as required by the local jurisdiction:

1. Do not allow the use of or operate spa if the suction outlet cover is missing, damaged or loose.
2. Check spa temperature before each use. Do not enter the spa if the temperature is above 104°F (40°C).
3. Keep breakable objects out of the spa area.
4. Spa shall not be operated during severe weather conditions.
5. Never place electrical appliances within 5 feet (1524 mm) of the spa.
6. No diving.

509.3 Depth markers. Public spas shall have permanent depth markers with numbers not less than 4 inches (102 mm) in height that are plainly and conspicuously visible from obvious points of entry and in conformance to this section.

509.3.1 Number. There shall be not less than two depth markers for each spa, regardless of spa size or shape.

509.3.2 Spacing. Depth markers shall be spaced at not more than 25-foot (7620 mm) intervals and shall be uniformly located around the perimeter of the spa.

509.3.3 Marking. Spas and exercise spas shall have the maximum water depth clearly marked on the required surfaces and such markers shall be positioned on the deck within 18 inches (457 mm) of the *design waterline*. Depth markers shall be positioned to be read while standing on the deck facing the water.

509.3.4 Slip resistant. Depth markers in or on the deck surfaces shall be slip resistant.

509.4 Clock. Public facilities shall have a clock that is visible to spa users.

CHAPTER 6
AQUATIC RECREATION FACILITIES

User note:

About this chapter: *Chapter 6 covers facilities commonly known as water parks. Such facilities can have a variety of pools ranging from zero depth entry wave pools to lazy rivers. This chapter includes requirements for floor slopes, steps, marking, signage, circulation systems, handholds and swimouts.*

SECTION 601 GENERAL

601.1 Scope. This chapter covers public pools and water containment systems used for aquatic recreation. This chapter provides specifications for the design, equipment, operation, signs, installation, sanitation, new construction, and rehabilitation of public pools for aquatic play. This chapter covers Class D-1 through Class D-6 public pools whether they are provided as stand-alone attractions or in various combinations in a composite attraction.

601.2 Combinations. Where combinations of Class D-1 through Class D-6 pools exist within a facility, each element in the facility shall comply with the applicable code sections as if the element functioned as a part of a freestanding pool of Class D-1 through Class D-6.

601.3 General. In addition to the requirements of this chapter, aquatic recreation facilities shall comply with the requirements of Chapter 3.

SECTION 602 FLOORS

602.1 Floor slope. In water depths of less than 5 feet (1524 mm), the floor slope shall be not greater than 1 unit vertical in 12 units horizontal (8.3-percent slope) except where the function of the attraction requires greater slopes in limited areas.

> **Exception:** The slope of the floor in Class D-3 pools shall not exceed 1 unit vertical in 7 units horizontal (14-percent slope).

SECTION 603 MARKINGS AND INDICATORS

603.1 Markings. Markings in areas of deep water shall comply with Section 409.2 except where the function of the pool dictates otherwise.

603.2 Class D-2 pools. Where a Class D-2 pool has a bather-accessible depth greater than $4^1/_2$ feet (1372 mm), the floor shall have a distinctive marking at the $4^1/_2$ feet (1372 mm) water depth.

603.3 Shallow-to-deep-end rope and float line. Where a pool has a water depth ranging from less than 5 feet (1524 mm) to greater than 5 feet (1524 mm), a rope and float line shall be located 1 foot (305 mm) horizontally from the 5-foot (1524 mm) depth location, toward the shallow end of the pool.

603.4 Nozzles. Pools having nonflush propulsion nozzles in the floor shall have a distinctive marking at the location of such nozzles.

SECTION 604 CIRCULATION SYSTEMS

604.1 General. A circulation system consisting of pumps, piping, return inlets and suction outlets, filters, and other necessary equipment shall be provided for complete circulation of water with the pool.

604.2 Turnover. Circulation system equipment shall be designed to turn over 100 percent of the nominal pool water volume in the amount of time specified in Table 604.2. The

TABLE 604.2 TURNOVER TIME

CLASS OF POOL	MAXIMUM TURNOVER TIME[a] (hours)
D-1	2
D-2 with less than 24 inches water depth	1
D-2 with 24 inches or greater water depth	2
D-3	1
D-4	2
D-5	1
D-6	1

For SI: 1 inch = 25.4 mm.

a. Pools with a sand bottom require a 1-hour turnover time.

system shall be designed to give the required turnover time based on the manufacturer's recommended maximum pressure and flow of the filter in clean media condition.

604.2.1 24-hour circulation required. Circulation systems shall circulate treated and filtered water for 24 hours a day.

604.2.2 Reduced circulation rate. The circulation rate shall be permitted to be reduced during periods that the pool is closed for use provided that acceptable water clarity conditions are met prior to reopening the pool for public use. The reduced circulation rate shall not be zero.

604.3 Surface skimming systems. Surface skimming systems shall be in accordance with Table 604.3.

604.3.1 Class D-5 pool skimmers. The installation of skimmers in the side areas of Class D-5 pools shall be prohibited.

SECTION 605 HANDHOLDS AND ROPES

605.1 Handholds. Handholds shall be provided in accordance with Section 323.

Exception: Handholds shall not be provided for wave action and action rivers.

605.2 Rope and float line. A rope and float line shall be provided for all of the following situations:

1. Separation of activity areas.
2. Identification of a break in floor slope at water depths of less than 5 feet (1524 mm).
3. Identification of a water depth greater than $4^{1}/_{2}$ feet (1372 mm) in constant floor slope in Class D-2 pools.

Exception: Class D-1 pools or any other pool where the designer indicates that such a line is not required or that the line would constitute a hazard.

605.2.1 Location. The rope and float line shall be located 1 foot (305 mm) toward the shallow end in each location.

605.3 Caisson wall rope and float line. For Class D-1 pools, a rope and float line shall be installed to restrict bather access to the wave pool caisson wall. The location of the rope and float line shall be in accordance with the wave equipment manufacturer's instructions.

605.4 Fastening. Rope and float lines shall be securely fastened to wall anchors made of corrosion-resistant materials. Wall anchors shall be of the recessed type and shall not have projections that will constitute a hazard when the rope and float line is removed.

605.5 Size. Rope and float lines shall be not less than $^{5}/_{8}$ inch (15.9 mm) in diameter and shall be made of polypropylene material.

SECTION 606 DEPTHS

606.1 Class D-6 depth. The captured or standing water depth in Class D-6 pools shall be not greater than 12 inches (305 mm).

606.2 Spray pools. The water depths in spray pools shall be not greater than 6 inches (152 mm).

SECTION 607 BARRIERS

607.1 Barriers. Multiple pools and spas within a single complex shall be permitted without barriers where a barrier separates the single complex from the surrounding property in accordance with Section 305.

SECTION 608 NUMBER OF OCCUPANTS

608.1 Occupant load. The occupant load for the pools or spas in the facility shall be calculated in accordance with Table 608.1. The occupant load shall be the combined total of the number of users based on the pool or spa water surface

TABLE 604.3 SURFACE SKIMMING SYSTEMS

CLASS OF POOL	SURFACE SKIMMING SYSTEM
D-1	Zero-depth trench located at static water level or other skimming systems
D-2	Auto skimmer, zero-depth trench or gutters
D-3	Auto skimmer, zero-depth trench or perimeter device
D-4	Single or multiple skimmer devices for skimming flow
D-5	Skimmers prohibited in side area
D-6	Auto skimmer, zero-depth trench, or gutter

TABLE 608.1 OCCUPANT LOAD

	SHALLOW OR WADING AREAS	DEEP AREA (NOT INCLUDING THE DIVING AREA)	DIVING AREA (PER EACH DIVING BOARD)	DECK AREA
Vessel water surface area	8 sq. ft. per user	10 sq. ft. per user	300 sq. ft. per user	—
Deck area	—	—	—	1 user per 15 sq. ft.

For SI: 1 square foot = 0.0929 m^2.

area and the deck area surrounding the pool or spa. The deck area occupant load shall be based on the occupant load calculated where a deck is provided or based on an assumed 4-foot-wide (1219 mm) deck surrounding the entire perimeter of the pool or spa, whichever is greater.

608.2 Facility capacity. For multiple pools and spas in a single aquatic recreation facility, the total facility occupant capacity shall not be limited by the number of occupants calculated in accordance with Section 608.1.

SECTION 609 TOILET ROOMS AND BATHROOMS

609.1 General. Toilet and bath facilities shall be in accordance with Sections 609.2 through 609.9.

609.2 Number of fixtures. Pools shall have toilet facilities with the number of fixtures in accordance with Section 609.2.1 or 609.2.2.

609.2.1 Water area less than 7500 square feet. Facilities that have less than 7500 gross square feet (697 m^2) of water area available for bather access shall have not less than one water closet for males, one urinal for males, one lavatory for males, one shower for males, two water closets for females, one lavatory for females and one shower for females.

609.2.2 Water area 7500 square feet or more. Facilities that have 7500 gross square feet (697 m^2) or more of water area available for bather access shall have not less than 0.7 water closet for males, one urinal for males, 0.85 lavatory for males, one shower for males, two water closets for females, one lavatory for females and one shower for females for every 7500 square feet (697 m^2) or portion thereof. Where the result of the fixture calculation is a portion of a whole number, the result shall be rounded up to the nearest whole number.

609.3 Showers. Showers shall be in accordance with Sections 609.3.1 through 609.3.5.

609.3.1 Deck shower. Not less than one and not more than half of the total number of showers required by Section 609.2 shall be located on the deck of or at the entrance of each pool.

609.3.2 Anti-scald device. Where heated water is provided to showers, the shower water supply shall be controlled by an anti-scald device.

609.3.3 Water heater and mixing valve. Bather access to water heaters and thermostatically controlled mixing valves for showers shall be prohibited.

609.3.4 Flow rate. Each showerhead shall have a water flow of not less than 2 gallons per minute (7.6 lpm).

609.3.5 Temperature. At each showerhead, the heated shower water temperature shall be not less than 90°F (32°C) and not greater than 120°F (49°C).

609.4 Soap dispensers. Soap dispensers shall be in accordance with Sections 609.4.1 and 609.4.2.

609.4.1 Liquid or powder. Soap dispensers shall be provided in each toilet facility. Soap dispensers shall dispense liquid or powdered soap. Reusable cake soap is prohibited.

609.4.2 Metal or plastic. Soap dispensers shall be made of metal or plastic. Glass materials shall be prohibited.

609.5 Toilet tissue holder. A toilet paper holder shall be provided at each water closet.

609.6 Lavatory mirror. Where mirrors are provided, they shall be shatter resistant.

609.7 Sanitary napkin receptacles. Sanitary napkin receptacles shall be provided in each water closet compartment for females and in the area of the showers for female use only.

609.8 Sanitary napkin dispensers. A sanitary napkin dispenser shall be provided in each toilet facility for females.

609.9 Infant care. Baby-changing tables shall be provided in toilet facilities having two or more water closets.

SECTION 610 SPECIAL FEATURES

610.1 Locations. Entry and exit locations shall be in accordance with Table 610.1. The primary means of entry and exit shall consist of ramps, beach entries, pool stairs, or ladders.

610.2 Secondary entry and exit means. Where secondary means of entry and exit are provided, they shall consist of one of the following:

1. Steps.
2. Stairs.
3. Ladders with grab rails.
4. Recessed treads.
5. Ramps.
6. Beach entries.
7. Swimouts.
8. Designs that provide the minimum utility as specified in this code.

610.3 Provisions for diving. Where diving facilities are part of the attraction or pool complex, entries, exits, pool stairs,

TABLE 610.1 ENTRY AND EXIT LOCATIONS

CLASS OF POOL	ENTRY AND EXIT LOCATIONS
D-1	Entry at beach end only; exit at beach end, sides or end wall
D-2	Entry and exit determined by the pool designer
D-3	Entry prohibited from deck areas; exit by ladders, steps or ramps as determined by pool designer
D-4	Entry and exit determined by the pool designer
D-5	Entry and exit determined by the pool designer
D-6	Entry and exit determined by the pool designer

ladders, underwater benches, special features, and other accessories shall be located outside of the minimum diving water envelope in accordance with Figure 322.2.

610.4 Beach entry, zero-depth entry, and sloping entries. The shallow end for beach entries and sloping entries shall be in accordance with Sections 610.4.1 through 610.4.4 or the regulations of the local jurisdiction.

610.4.1 Maximum entry slope. The slope of sloping entries used as a pool entry shall not exceed 1 unit vertical in 12 units horizontal (8.3-percent slope).

610.4.2 Benches. Where benches are used in conjunction with sloping entries, the vertical riser height shall not exceed 12 inches (305 mm).

610.4.3 Steps. Where steps are used in conjunction with sloping entries, all of the requirements of Section 610.5 shall apply.

610.4.4 Slip-resistant surfaces. Beach and sloping entry walking surfaces at water depths up to 18 inches (457 mm) shall be slip resistant.

610.5 Pool steps. The design and construction of steps for stairs into the shallow end and recessed pool stairs shall be in accordance with Sections 610.5.1 through 610.5.6.

610.5.1 Uniform height of 9 inches. Except for the bottom riser, risers at the centerline shall have a maximum uniform height of 9 inches (229 mm). The bottom riser height shall be permitted to vary from the other risers.

610.5.2 Distance from coping or deck. The vertical distance from the pool coping, deck, or step surface to the uppermost tread shall be not greater than 9 inches (229 mm).

610.5.3 Color to mark leading edge. The leading edge of steps shall be distinguished by a color contrasting with the color of the steps and the pool floor.

610.5.4 Stairs in water depths over 48 inches. Stairs that are located in water depths greater than 48 inches (1219 mm) shall have the lowest tread located below the deck at a distance of not less than 48 inches (1219 mm) below the deck.

610.5.5 Tread horizontal depth. Treads shall have an unobstructed horizontal depth of not less than 11 inches (279 mm).

610.5.6 Tread surface area. Treads shall have an unobstructed surface area of not less than 240 square inches (.017 m^2).

610.6 Swimouts. Swimouts shall be located completely outside of the water current or wave action of the pool or spa and can be located in shallow or deep areas of water.

610.6.1 Surface area. An unobstructed surface equal to or greater than that required for the top tread of the pool stairs shall be provided in accordance with Sections 610.5.5 and 610.5.6.

610.6.2 Step required. Where a swimout is used as an entry and exit access point, it shall be provided with a step that meets the pool stair requirements (see Section 610.5).

610.6.3 Maximum depth. The horizontal surface of a swimout shall be not greater than 20 inches (508 mm) below the waterline.

610.6.4 Color marking. The leading edge of a swimout shall be visually set apart by a stripe having a width of not less than $^3/_4$ inch (19 mm) and not greater than 2 inches (51 mm). The stripe shall be of a contrasting color to the adjacent surfaces.

610.7 Underwater seats and benches. Underwater seats and benches shall comply with this section.

610.7.1 Prohibited location. Underwater seats shall not be located in the diving water envelope.

610.7.2 Surface dimensions. Underwater seats shall have an unobstructed surface dimension of not less than 10 inches (254 mm) measured front to back and not less than 24 inches (610 mm) in width.

610.7.3 Not an entry or exit. Underwater seats and benches shall not be used as an entry or exit for a pool but can be located in shallow or deep areas of water.

610.7.4 Depth. The horizontal surface of seats and benches shall be not greater than 20 inches (508 mm) below the waterline.

610.7.5 Color marking. The leading edge of seats and benches shall be visually set apart by a stripe having a width not less than $^3/_4$ inch (19 mm) and not greater than 2 inches (51 mm). The stripe shall be of a contrasting color to the adjacent surfaces.

610.7.6 Slip resistant. The top surface of seats and benches shall be slip resistant.

610.8 Objects permitted. The design, construction, and operation of decorative objects and structures intended for climbing, walking, and hanging on by a bather are not covered by this code.

610.8.1 Floating devices. Floating devices not intended to be mobile shall be anchored in a manner to restrict movement to the range established by the designer. The anchoring of such floating devices shall be configured to minimize the possibility of entrapment of bathers, bodies, hair, limbs, and appendages should they come in contact with any element of the floating device or its anchors.

SECTION 611
SIGNAGE

611.1 Posting of signs. Signs stating rules, instructions, and warnings shall be posted. Signs for suction entrapment warning in accordance with Section 310 shall be posted. Signs shall be placed so that they squarely face approaching traffic. The center of the message panel shall be located not less than 66 inches (1676 mm) above the walking surface.

611.2 Prohibited mounting. Signs shall not be mounted on fences and gates alongside of guest walkways and staircases.

611.3 Message delivery. Messages delivered on signs shall comply with all of the following:

1. Messages shall be pertinent to the activity being performed or to be performed.

2. Messages shall be specific by providing details about the activity.
3. Messages shall be short and concise.
4. Messages shall be direct without humor or embellishments.

611.4 Text font and size. The message text shall be in a clear, bold font such as Arial. The character height shall be proportional to 1 inch (25 mm) for 10 feet (3048 mm) of intended viewing distance but not less than 1 inch (25 mm).

611.5 Distinct sign classes. Facility signs shall be categorized into four sign classes in accordance with Sections 611.5.1 through 611.5.4.

611.5.1 General information. General information signs shall be posted facility-wide and shall not be attraction specific.

611.5.2 Directional signs. Directional signs shall identify the location of services and attractions in the park and shall include directional arrows. Directional signs shall be posted at various crossroads in the facility.

611.5.3 Rule signs. Rule signs shall inform guests of the qualifications that they must meet to allow them to participate on a specific ride or attraction. Rules shall include, but are not limited to, limits for weight and height, proper attire and ride (and ride vehicle) stipulations. Rule signs shall be located at a point where the guests make the initial commitment to participate on the ride.

611.5.4 Instructional signs. Instructional signs shall inform guests of specific instructions for the use of the ride. Instructions shall include, but are not limited to, riding posture, prohibited activity, and user exit requirements at the ride termination. Instructional signs shall be located along the queue approaching the ride dispatch area.

611.6 Materials. Sign panels shall be durable for the weather conditions and shall be resistant to damage from guests. The message surface shall be clean and smooth and shall readily accept paint or precut lettering adhesives.

611.7 Shape and size consistency. The panel shape and size for each class of signs shall be the same. Where the total message to be indicated is larger than what can be placed on one sign, multiple signs of the same size shall be used to display the message.

611.8 Pictograms. Pictograms shall always be accompanied by text indicating the same message. Pictograms shall be designed to illustrate one clear and specific meaning to all individuals.

611.9 Theming or artwork. Theming or artwork applied to signs shall not invade the message panel. Signs shall have a distinct border.

611.10 Shallow water. Safety signs shall be in accordance with Section 412.

611.11 Cold water. Where a pool could have a water temperature below 70°F (21°C), a cold water warning sign shall be posted at the point of entry to the pool or at the attraction using such water.

CHAPTER 7

ONGROUND STORABLE RESIDENTIAL SWIMMING POOLS

User note:

About this chapter: *Chapter 7 concerns residential portable pools also known as onground storable residential swimming pools. These pools are manufactured for assembly on the site. The chapter's regulations include those for floor slopes, entry barrier methods, decks, stairs, safety signage and circulation systems.*

SECTION 701
GENERAL

701.1 Scope. This chapter describes certain criteria for the design, manufacturing, and testing of *onground storable pools* intended for *residential* use. This includes portable pools with flexible or nonrigid side walls that achieve their structural integrity by means of uniform shape, support frame or a combination thereof, and that can be disassembled for storage or relocation. This chapter includes what has been commonly referred to in past standards or codes as onground or above-ground pools.

701.1.1 Permanent inground residential swimming pool. This chapter does not apply to permanent inground *residential* pools, as defined in Chapter 8.

701.2 General. In addition to the requirements of this chapter, onground storable *residential* swimming pools shall comply with the requirements of Chapter 3.

701.3 Floor slopes. Floor slopes shall be uniform and in accordance with Sections 701.3.1 through 701.3.4.

701.3.1 Shallow end. The slope of the floor from the shallow end wall towards the deep area shall not exceed 1 unit vertical in 7 units horizontal (14-percent slope) to the point of the first slope change.

701.3.2 Transition. The slope of the floor from the point of the first slope change towards the deepest point shall not exceed 1 unit vertical in 3 units horizontal (33-percent slope).

701.3.3 Adjacent. The slope adjacent to the shallow area shall not exceed 1 unit vertical in 3 units horizontal (33-percent slope) and the slope adjacent to the side walls shall not exceed 1 unit vertical in 1 unit horizontal (100-percent slope).

701.3.4 Change point. The point of the first slope change shall be defined as the point at which the shallow area slope exceeds 1 unit vertical in 7 units horizontal (14-percent slope) and is not less than 6 feet (1889 mm) from the shallow end wall of the pool.

701.4 Identification. For onground storable *residential* pools with a vinyl liner, the manufacturer's name and the liner identification number shall be affixed to the liner. For onground storable *residential* pools without a liner, the manufacturer's name and identification number shall be affixed to the exterior of the pool structure.

701.5 Installation. *Onground storable pools* shall be installed in accordance with the manufacturer's instructions.

SECTION 702
LADDERS AND STAIRS

702.1 Ladders and stairs. Pools shall have a means of entry and exit consisting of not less than one ladder or a ladder and staircase combination.

702.2 Type A and Type B ladders. Type A, double access, and Type B, limited access, A-frame ladders shall comply with Sections 702.2.1 through 702.2.7. See Figure 702.2.

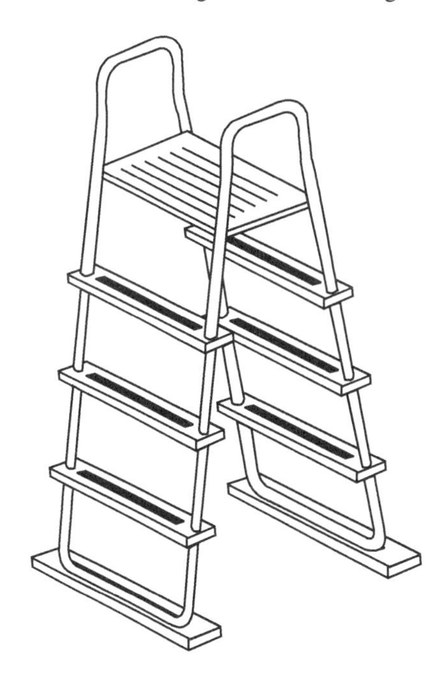

FIGURE 702.2
TYPICAL A-FRAME LADDER, TYPES A AND B

702.2.1 Barrier required. Ladders in the pool shall have a physical barrier to prevent children from swimming through the riser openings or behind the ladder.

Exception: Barriers for ladders shall not be required where the ladder manufacturer provides a certification statement that the ladder complies with the ladder entrapment test requirements of APSP 4.

702.2.2 Platform. Where an A-frame ladder has a platform between the handrails, the platform shall have a width of not less than 12 inches (305 mm) and a length of not less than 12 inches (305 mm). The platform shall be at or above the highest ladder tread. The walking surface of the platform shall be slip resistant.

702.2.3 Handrails or handholds. A-frame ladders shall have two handrails or handholds that serve all treads. The height of the handrails and handholds shall be not less than 20 inches (508 mm) above the platform or uppermost tread, whichever is higher.

702.2.4 Diameter. The outside diameter of handrails and handholds shall be not less than 1 inch (25 mm) and not greater than 1.9 inches (48 mm).

702.2.5 Clear distance. The clear distance between ladder handrails shall be not less than a space of 12 inches (305 mm).

702.2.6 Treads. Ladder treads shall have a horizontal uniform depth of not less than 2 inches (51 mm).

702.2.7 Riser height. Risers, other than the bottom riser, shall be of uniform height that is not less than 7 inches (178 mm) and not greater than 12 inches (305 mm). The bottom riser height shall be not less than 7 inches (178 mm) and not greater than 12 inches (305 mm). The vertical distance from the platform or top of the pool structure to the uppermost tread shall be the same as the uniform riser heights.

702.3 Type C staircase ladders (ground to deck). Type C staircase ladders shall comply with Sections 702.3.1 through 702.3.6. See Figure 702.3.

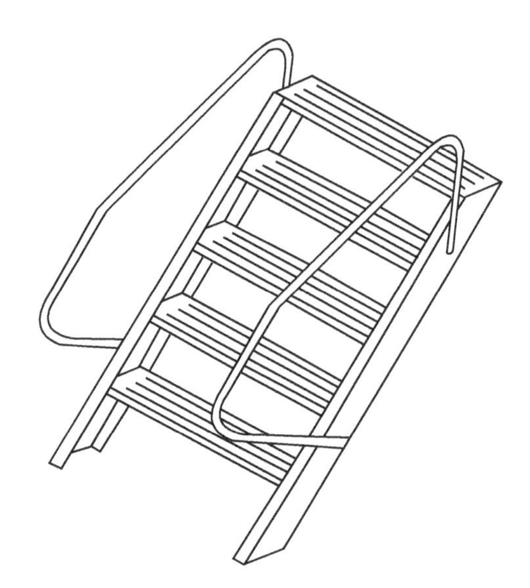

FIGURE 702.3
TYPICAL STAIRCASE LADDER, TYPE C

702.3.1 Handrails or handholds. Staircase ladders shall have not less than two handrails or handholds that serve all treads. The height of the handrails and handholds shall be not less than 20 inches (508 mm) above the platform or uppermost tread, whichever is higher.

702.3.2 Diameter. The outside diameter of handrails and handholds shall be not less than 1 inch (25 mm) and not greater than 1.9 inches (48 mm).

702.3.3 Treads. Ladder treads shall have a horizontal uniform depth of not less than 4 inches (102 mm).

702.3.4 Riser height. Risers, other than the bottom riser, shall be of uniform height that is not less than 7 inches (178 mm) and not greater than 12 inches (305 mm). The bottom riser height shall be not less than 7 inches (178 mm) and not greater than 12 inches (305 mm). The vertical distance from the platform or top of the pool structure to the uppermost tread shall be the same as the uniform riser heights.

702.3.5 Top step. The top step of a staircase ladder shall be flush with the deck or 7 inches (178 mm) to 12 inches (305 mm) below the deck level.

702.3.6 Width. Steps shall have a minimum unobstructed width of 19 inches (483 mm) between the side rails.

702.4 Type D in-pool ladders. Type D in-pool ladders shall be in accordance with Sections 702.4.1 through 702.4.7. See Figure 702.4.

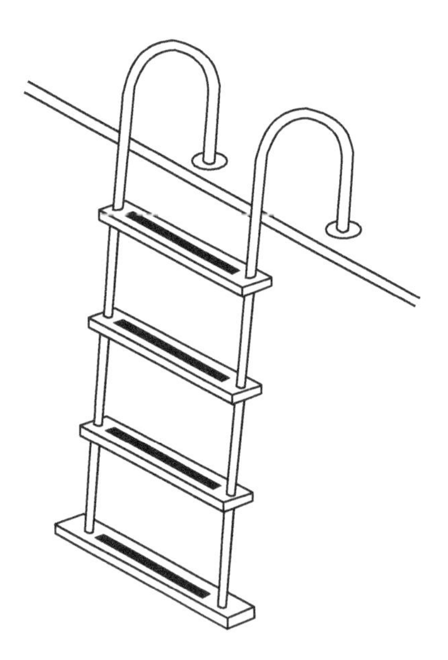

FIGURE 702.4
TYPICAL IN-POOL LADDER, TYPE D

702.4.1 Clearance. There shall be a clearance of not less than 3 inches (76 mm) and not greater than 6 inches (152 mm) between the pool wall and the ladder.

702.4.2 Handrails or handholds. Ladders shall be equipped with two handrails or handholds that extend above the platform or deck not less than 20 inches (508 mm).

702.4.3 Clear distance. The clear distance between ladder handrails shall be not less than 12 inches (305 mm).

702.4.4 Diameter. The outside diameter of handrails and handholds shall be not less than 1 inch (25 mm) and not greater than 1.9 inches (48 mm).

702.4.5 Riser height. Risers, other than the bottom riser, shall be of uniform height that is not less than 7 inches (178 mm) and not greater than 12 inches (305 mm). The bottom riser height shall be not less than 7 inches (178 mm) and not greater than 12 inches (305 mm).

702.4.6 Top tread. The vertical distance from the pool coping, deck, or step surface to the uppermost tread shall be not less than 7 inches (178 mm) and not greater than 12 inches (305 mm) and uniform with other riser heights.

702.4.7 Tread depth. Ladder treads shall have a horizontal uniform depth of not less than 2 inches (51 mm).

702.5 Type E protruding in-pool stairs. Type E protruding in-pool stairs shall be in accordance with Sections 702.5.1 through 702.5.7. See Figure 702.5.

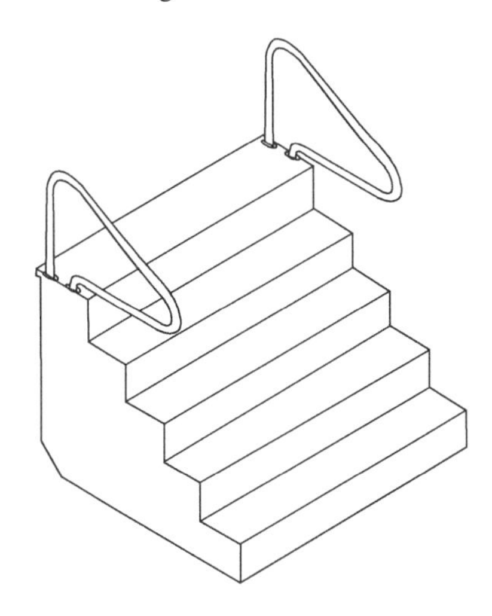

FIGURE 702.5
TYPICAL IN-POOL STAIRCASE, TYPES E AND F

702.5.1 Barrier required. In-pool stairs shall have a physical barrier to prevent children from swimming through the riser openings or behind the in-pool stairs.

702.5.2 Handrails or handholds. In-pool stairs shall be equipped with not less than one handrail or handhold that serves all treads with a height of not less than 20 inches (508 mm) above the platform or uppermost tread, whichever is higher.

702.5.3 Removable handrails. Where handrails are removable, they shall be installed such that they cannot be removed without the use of tools.

702.5.4 Leading edge distance. The leading edge of handrails shall be 18 inches (457 mm) ± 3 inches (± 76 mm), horizontally from the vertical plane of the bottom riser.

702.5.5 Diameter. The outside diameter of handrails and handholds shall be not less than 1 inch (25 mm) and not greater than 1.9 inches (48 mm).

702.5.6 Tread width and depth. Treads shall have an unobstructed horizontal depth of not less than 10 inches (254 mm) and an unobstructed surface area of not less than 240 square inches (0.15 m^2).

702.5.7 Uniform riser height. Risers, other than the bottom riser, shall be of uniform height that is not less than 7 inches (178 mm) and not greater than 12 inches (305 mm). The bottom riser height shall be not less than 7 inches (178 mm) and not greater than 12 inches (305 mm). The vertical distance from the pool coping, deck or step surface to the uppermost tread of the stairs shall be the same as the uniform riser heights.

702.6 Type F recessed in-pool stairs. Type F recessed in-pool stairs shall be in accordance with Sections 702.6.1 through 702.6.7. See Figure 702.5.

702.6.1 Barrier required. In-pool stairs shall have a physical barrier to prevent children from swimming through the riser openings or behind the in-pool stairs.

702.6.2 Handrails or handholds. In-pool stairs shall be equipped with not less than one handrail or handhold that serves all treads with a height of not less than 20 inches (508 mm) above the platform or uppermost tread, whichever is higher.

702.6.3 Removable handrails. Where handrails are removable, they shall be installed such that they cannot be removed without the use of tools.

702.6.4 Leading edge distance. The leading edge of handrails shall be 18 inches (457 mm) ± 3 inches (± 76 mm), horizontally from the vertical plane of the bottom riser.

702.6.5 Diameter. The outside diameter of handrails and handholds shall be not less than 1 inch (25 mm) and not greater than 1.9 inches (48 mm).

702.6.6 Tread width and depth. Treads shall have an unobstructed horizontal depth of not less than 10 inches (254 mm) at all points and an unobstructed surface area of not less than 240 square inches (0.15 m^2).

702.6.7 Uniform riser height. Risers, other than the bottom riser, shall be of uniform height that is not less than 7 inches (178 mm) and not greater than 12 inches (305 mm). The bottom riser height shall be not less than 7 inches (178 mm) and not greater than 12 inches (305 mm). The vertical distance from the pool coping, deck or step surface to the uppermost tread of the stairs shall be the same as the uniform riser heights.

SECTION 703
DECKS

703.1 General. Decks provided by the pool manufacturer shall be installed in accordance with the manufacturer's instructions. Decks fabricated on-site shall be in accordance with the *International Residential Code*.

703.2 Cantilevered. The top surface of a cantilevered deck shall be not greater than 1 inch (25 mm) higher than the top of the pool wall. See Figure 703.4. The top surface of a noncantilevered deck shall be not higher than the top of the pool wall.

703.3 No gaps. Decks that are installed flush with the top rail of the pool shall have all gap openings between the deck and top rails closed-off or capped.

703.4 Extension over pool. Where a deck extends inside the top rail of the pool, it shall extend not more than 3 inches (76 mm) beyond the inside of the top rail of the pool in accordance with Figure 703.4 and shall have a smooth finish.

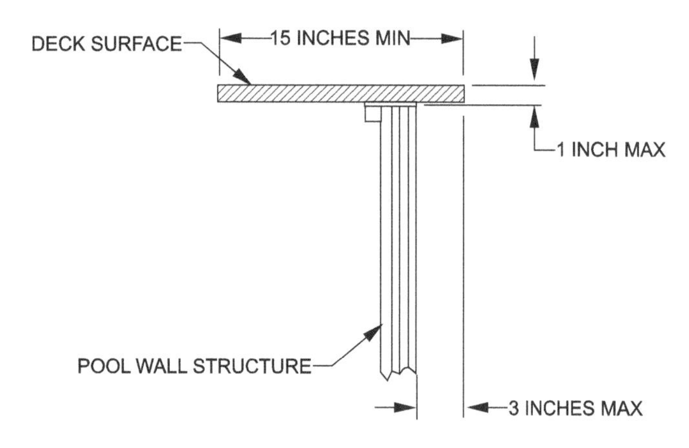

For SI: 1 inch = 25.4 mm.

FIGURE 703.4
TYPICAL CANTILEVERED DECK SUPPORT

703.5 Slip resistant. The deck walking surface shall be slip resistant.

703.6 Walk-around decks. Walk-around decks shall have a level walking surface of not less than 15 inches (381 mm) in width, as measured from the inside edge of the pool top rail to the outside of the pool walk-around. See Figure 703.6.

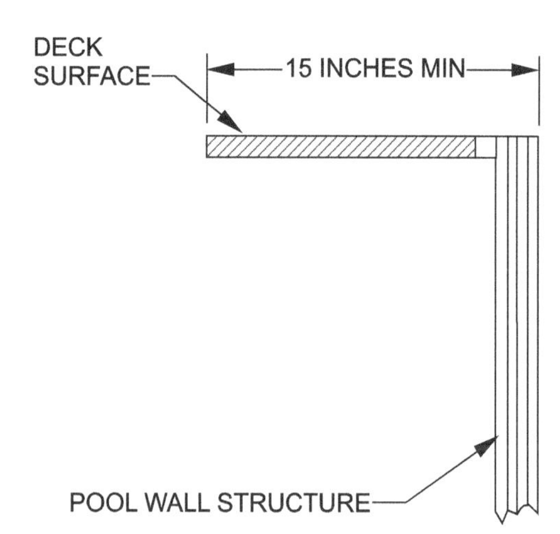

For SI: 1 inch = 25.4 mm.

FIGURE 703.6
WALK-AROUND DECK WIDTH

SECTION 704
CIRCULATION SYSTEM

704.1 General. A circulation system consisting of pumps, hoses, tubing, piping, return inlets, suction outlets, filters and other related equipment that provides for the circulation of water throughout the pool shall be located so that such items cannot be used by young children as a means of access to the pool.

704.2 Installation and support. Circulation equipment shall be installed, mounted and supported in accordance with the manufacturer's instructions.

704.3 Draining the system. In climates subject to freezing, circulation system equipment shall be designed and fabricated to drain the pool water from the equipment and exposed piping, by removal of drain plugs and manipulating valves or by other methods in accordance with the manufacturer's instructions.

704.4 Turnover. A pump including a motor shall be provided for circulation of the pool water. The equipment shall be sized to provide a turnover of the pool water not less than once every 12 hours. The system shall be designed to provide the required turnover rate based on the manufacturer's specified maximum flow rate of the filter, with a clean media condition of the filter. The system flow shall not exceed the filter manufacturer's maximum filter flow rate.

704.5 Piping and fittings. The process piping of the circulation system, including but not limited to hoses, tubing, piping, and fittings, shall be made of nontoxic material and shall be capable of withstanding an internal pressure of not less than $1^1/_2$ times the rated pressure of the pump. Piping on the suction side of the pump shall not collapse when flow into such piping is blocked.

704.6 Filters. Pressure-type filters shall have an automatic internal means or a manual external means to relieve accumulated air pressure inside the filter tank. Filter tanks composed of upper and lower tank lids that are held in place by a perimeter clamp shall have a perimeter clamp that provides for a slow and safe release of air pressure before the clamp disengages the lids.

704.6.1 Automatic internal air relief. Filter tanks incorporating an automatic internal air relief as the principal means of air release shall be designed with a means to provide for a slow and safe release of pressure.

704.6.2 Separation tank. A separation tank used in conjunction with a filter tank shall have a manual air release or the tank shall be designed to provide for a slow and safe release of pressure when the tank is opened.

704.7 Pumps. Pool pumps shall be tested and certified by a nationally recognized testing laboratory in accordance with UL 1081.

704.7.1 Cleanable strainer. Where a pressure-type filter is installed, a cleanable strainer or screen that captures materials such as solids, debris, hair and lint shall be provided upstream of the circulation pump.

704.7.2 Accessible pumps and motors. Pumps and motors shall be accessible for inspection and service in accordance with the pump and motor manufacturer's instructions.

704.7.3 Pump shutoff valves. An *accessible* means of shut off of the suction and discharge piping for the pump shall be provided for maintenance and removal of the pump.

704.8 Outlets and return inlets. Outlets or suction outlets and return inlets shall be provided and arranged to produce uniform circulation of water so that sanitizer residual is maintained throughout the pool. Where installed, submerged suction outlets shall conform to APSP 16.

704.9 Surface skimmer systems. The surface skimming system provided shall be designed and constructed to skim the pool surface where the water level is maintained between the minimum and maximum fill level of the pool.

704.9.1 Coverage where used as a sole outlet. Where surface skimmers are used as the only pool water outlet system, not less than one skimmer shall be provided for each 800 square feet (74.3 m^2), or fraction thereof, of the water surface area.

704.9.2 Coverage where used in combination with other outlets. Where surface skimmers are not the only outlet for pool water, they shall be considered to cover only that fraction of the 800 square feet (74.3 m^2).

704.9.3 Location and venting. Skimmers shall be equipped with a vent that serves as a vacuum break.

SECTION 705 SAFETY SIGNS

705.1. Signs to be installed prior to final inspection. Safety signage such as "NO DIVING" signs and other safe use instruction signs that are provided by the pool and ladder manufacturer shall be posted in accordance with the manufacturer's instructions prior to final inspection.

705.2 Safety signs for ladders. Safety signage for ladders shall be in accordance with Sections 705.2.1 through 705.2.3.2.

705.2.1 A-frame ladders. Safety signage for A-frame ladders shall be in accordance with Sections 705.2.1.1 through 705.2.1.4.1. The words on the signage shall be readable by persons standing in the pool and standing outside of the pool as applicable for the required location of each sign.

705.2.1.1 No diving warning. A-frame ladders shall have the following words posted on the in-pool side of the ladder and on the pool entry side of the ladder: "NO DIVING." The location of the words shall be above the elevation of the design water level of the pool.

705.2.1.2 Entrapment warning. A-frame ladders shall have the following words posted on the pool side of the ladder: "TO PREVENT ENTRAPMENT OR DROWNING DO NOT SWIM THROUGH, BEHIND, OR AROUND LADDER."

705.2.1.3 Type A, A-frame ladders. Type A double access A-frame ladders shall have the following words posted on the ladder: "REMOVE AND SECURE LADDER WHEN POOL IS NOT OCCUPIED."

705.2.1.4 Type B, A-frame ladders. Type B limited access A-frame ladders shall have the following words posted on the ladder: "SECURE LADDER WHEN POOL IS NOT OCCUPIED."

705.2.1.4.1 Swing up or slide up secured ladders. Type B limited access A-frame ladders that utilize swing-up or slide-up sections for limiting access to the pool shall have the following words posted on the ladder as applicable for the type of securing method:

1. "WHEN POOL IS NOT OCCUPIED, SWING UP AND SECURE."
2. "WHEN POOL IS NOT OCCUPIED, LIFT OFF."
3. "WHEN POOL IS NOT OCCUPIED, SLIDE UP AND SECURE."

705.2.2 Type C staircase ladders. Type C staircase ladders that swing up to limit access to the pool or that are removed to limit access to the pool shall have the following words posted on the ladder: "WHEN NOT IN USE SWING UP AND SECURE OR REMOVE."

705.2.3 Type D in-pool ladder. Safety signage for Type D in-pool ladders shall be in accordance with Sections 705.2.3.1 and 705.2.3.2. The words on the signage shall be readable by persons standing in the pool or standing outside the pool as applicable for the required location of each sign.

705.2.3.1 No diving warning. Type D in-pool ladders shall have the following words posted on the in-pool side of the ladder and on the pool entry side of the ladder: "NO DIVING." The location of the words shall be above the elevation of the design water level of the pool.

705.2.3.2 Entrapment warning. Type D in-pool ladders shall have the following words posted on the ladder: "WARNING: TO PREVENT ENTRAPMENT OR DROWNING, DO NOT SWIM THROUGH, BEHIND, OR AROUND LADDER."

CHAPTER 8

PERMANENT INGROUND RESIDENTIAL SWIMMING POOLS

User note:

About this chapter: *Permanent inground residential swimming pools are regulated by Chapter 8. Where diving boards are present, this chapter provides information regarding the minimum diving water dimensions. Requirements for means of entry and exit, decks and circulation systems are provided. Special features of these pools such as beach entries, swimouts, diving rocks and architectural features are also regulated by this chapter.*

SECTION 801 GENERAL

801.1 Scope. The provisions of this chapter shall govern permanent inground *residential* swimming pools. Permanent inground *residential* swimming pools shall include pools that are partially or entirely above grade. This chapter does not cover pools that are specifically manufactured for aboveground use and that are capable of being disassembled and stored. This chapter covers new construction, modification and repair of inground *residential* swimming pools.

801.2 General. Permanent inground *residential* pools shall comply with the requirements of Chapter 3.

SECTION 802 DESIGN

802.1 Materials of components and accessories. The materials of components and accessories used for permanent inground *residential* swimming pools shall be suitable for the environment in which they are installed. The materials shall be capable of fulfilling the design, installation and the intended use requirements in the *International Residential Code*.

802.2 Structural design. The structural design and materials shall be in accordance with the *International Residential Code*.

SECTION 803 CONSTRUCTION TOLERANCES

803.1 Construction tolerances. The construction tolerance for dimensions for the overall length, width and depth of the pool shall be ± 3 inches (76 mm). The construction tolerance for all other dimensions shall be ± 2 inches (51 mm), unless otherwise specified by the design engineer.

SECTION 804 DIVING WATER ENVELOPES

804.1 General. The minimum diving water envelopes shall be in accordance with Table 804.1 and Figure 804.1. Negative construction tolerances shall not be applied to the dimensions of the minimum diving water envelopes given in Table 804.1.

SECTION 805 WALLS

805.1 General. Walls in the shallow area and deep area of the pool shall have a wall-to-floor transition point that is not less than 33 inches (838 mm) below the *design waterline*. Above the transition point, the walls shall be within 11 degrees (0.19 rad) of vertical.

SECTION 806 OFFSET LEDGES

806.1 Maximum width. Offset ledges shall be not greater than 8 inches (203 mm) in width.

806.2 Reduced width required. Where an offset ledge is located less than 42 inches (1067 mm) below the *design waterline*, the width of such ledge shall be proportionately less than 8 inches (203 mm) in width so as to fall within 11 degrees of vertical as measured from the top of the *design waterline*.

TABLE 804.1
MININUM DIVING WATER ENVELOPE FOR SWIMMING POOLS DESIGNATED TYPES I-V[b]

POOL TYPE	MINIMUM DEPTHS AT POINT FEET-INCHES				MINIMUM WIDTHS AT POINT FEET-INCHES				MINIMUM LENGTHS BETWEEN POINTS FEET-INCHES					
	A	B	C	D	A	B	C	D	WA	AB	BC	CD	DE	WE
I	6-0	7-6	5-0	2-9	10-0	12-0	10-0	8-0	1-6	7-0	7-6	Note a	6-0	28-9
II	6-0	7-6	5-0	2-9	12-0	15-0	12-0	8-0	1-6	7-0	7-6	Note a	6-0	28-9
III	6-10	8-0	5-0	2-9	12-0	15-0	12-0	8-0	2-0	7-6	9-0	Note a	6-0	31-3
IV	7-8	8-0	5-0	2-9	15-0	18-0	15-0	9-0	2-6	8-0	10-6	Note a	6-0	31-3
V	8-6	9-0	5-0	2-9	15-0	18-0	15-0	9-0	3-0	9-0	12-0	Note a	6-0	36-9

For SI: 1 inch = 25.4 mm, 1 foot = 304.8 mm.

a. The minimum length between points C and D varies based on water depth at point D and the floor slope between points C and D.

b. See Figure 804.1 for location of points.

SECTION 807
POOL FLOORS

807.1 Floor slopes. Floor slopes shall be in accordance with Sections 807.1.1 through 807.1.3.

807.1.1 Shallow end. The slope of the floor from the beginning of the shallow end to the deep area floor slope transition point, indicated in Figure 804.1 as Point E to Point D, shall not exceed 1 unit vertical in 7 units horizontal.

807.1.2 Shallow to deep transition. The shallow to deep area floor slope transition point, indicated in Figure 804.1 as Point D, shall occur at a depth not less than 33 inches (838 mm) below the *design waterline* and at a point not less than 6 feet (1829 mm) from the beginning of the shallow end, indicated in Figure 804.1 as Point E, except as specified in Section 809.7.

807.1.3 Deep end. The slope of the floor in the deep end, indicated in Figure 804.1 as Point B to Point D, shall not exceed a slope of 1 unit vertical in 3 units horizontal (33-percent slope).

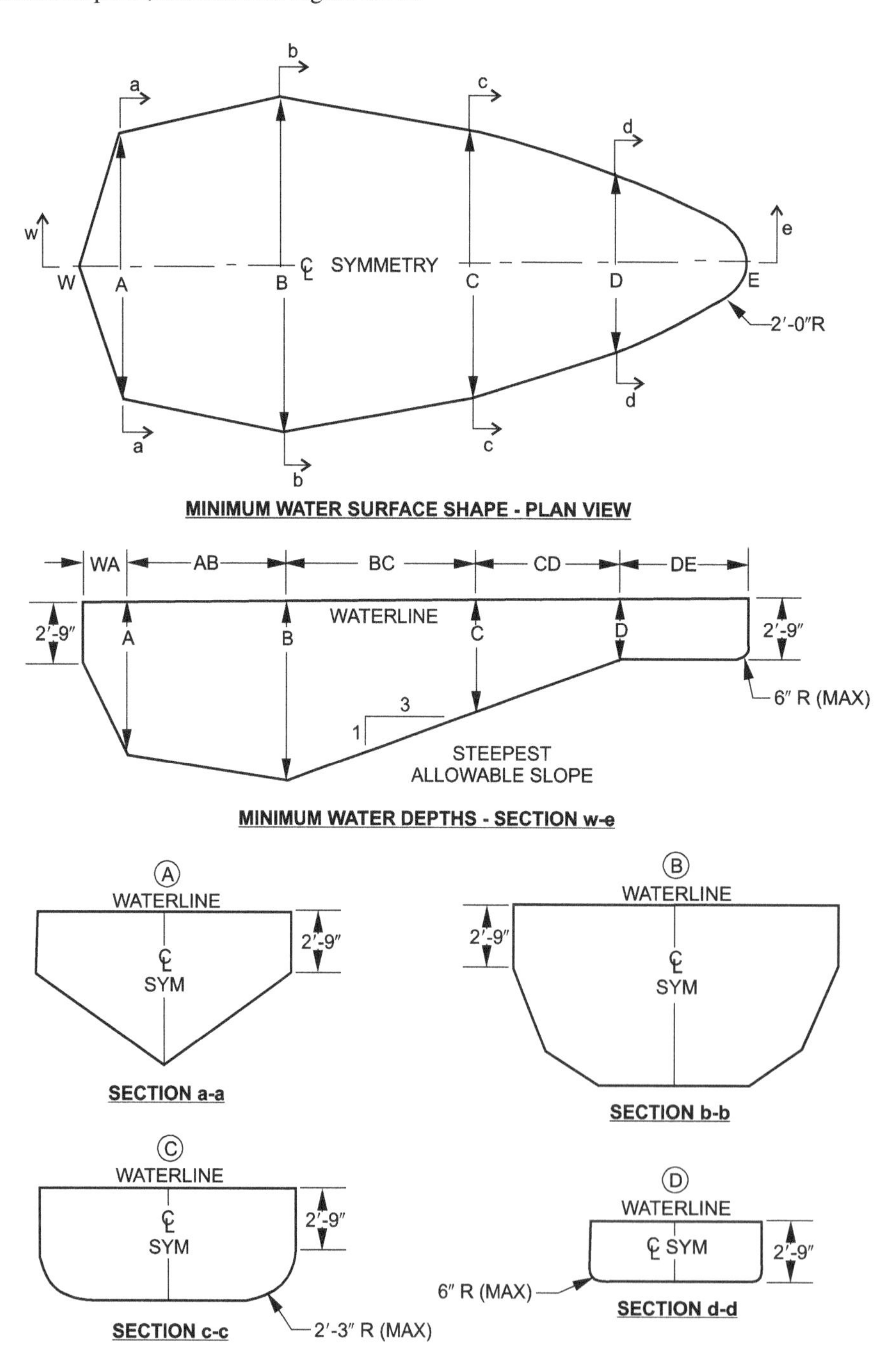

For SI: 1 inch = 25.4 mm, 1 foot = 304.8 mm.

FIGURE 804.1
MINIMUM DIVING WATER ENVELOPE

807.2 Shallow end water depths. The design water depth as measured at the shallowest point in the shallow area shall be not less than 33 inches (838 mm) and not greater than 4 feet (1219 mm). Shallow areas designed in accordance with Sections 809.6, 809.7 and 809.8 shall be exempt from the minimum depth requirement.

SECTION 808
DIVING EQUIPMENT

808.1 Manufactured and fabricated diving equipment. Manufactured and fabricated diving equipment shall be in accordance with this section. Manufactured and fabricated diving equipment and appurtenances shall not be installed on a Type O pool.

808.2 Manufactured diving equipment. Manufactured diving equipment shall be designed for swimming pool use.

808.3 Installation. Where manufactured diving equipment is installed, the installation shall be located in the deep area of the pool so as to provide the minimum dimensions as shown in Table 804.1 and shall be installed in accordance with the manufacturer's instructions.

808.4 Labeling. Manufactured diving equipment shall have a permanently affixed label indicating the manufacturer's name and address, the date of manufacture, the minimum diving envelope and the maximum weight limitation.

808.5 Slip resistant. Diving equipment shall have slip-resistant walking surfaces.

808.6 Point A. For the application of Table 804.1, Point A shall be the point from which all dimensions of width, length and depth are established for the minimum diving water envelope. If the tip of the diving board or diving platform is located at a distance of WA or greater from the deep end wall and the water depth at that location is equal to or greater than the water depth requirement at Point A, then the point on the water surface directly below the center of the tip of the diving board or diving platform shall be identified as Point A.

808.7 Location of pool features in a diving pool. Where a pool is designed for use with diving equipment, the location of steps, pool stairs, ladders, underwater benches, special features and other accessory items shall be outside of the minimum diving water envelope as indicated in Figure 322.2.

808.8 Stationary diving platforms and diving rocks. Stationary diving platforms and diving rocks built on-site shall be permitted to be flush with the wall and shall be located in the diving area of the pool. Point A shall be in front of the wall at the platform or diving rock centerline.

808.9 Location. The forward tip of manufactured or fabricated diving equipment shall be located directly above Point A as defined by Section 808.6.

808.10 Elevation. The maximum elevation of a diving board above the *design waterline* shall be in accordance with the manufacturer's instructions.

808.11 Minimum water envelope. Manufactured diving equipment installation and use instructions shall be provided by the diving equipment manufacturer and shall specify the minimum water dimensions required for each diving board and diving stand combination. The board manufacturer shall indicate the water envelope type by dimensionally relating their products to Point A on the water envelopes as shown in Figure 804.1 and Table 804.1. The board manufacturer shall specify which boards fit on the design pool geometry types as indicated in Table 804.1.

808.12 Platform height above waterline. The height of a stationary diving platform or a diving rock above the *design waterline* shall not exceed the dimensions in Table 808.12.

TABLE 808.12
DIVING PLATFORM OR APPURTENANCE HEIGHT ABOVE DESIGN WATERLINE

POOL TYPE	HEIGHT INCHES
I	42
II	42
III	50
IV	60
V	69

For SI: 1 inch = 25.4 mm.

808.13 Headroom above the board. The diving equipment manufacturer shall specify the minimum headroom required above the board tip.

SECTION 809
SPECIAL FEATURES

809.1 Slides. Slides shall be installed in accordance with the manufacturer's instructions.

809.2 Entry and exit. Pools shall have a means of entry and exit in all shallow areas where the design water depth of the shallow area at the shallowest point exceeds 24 inches (610 mm). Entries and exits shall consist of one or a combination of the following: steps, stairs, ladders, treads, ramps, beach entries, underwater seats, benches, swimouts and other *approved* designs. The means of entry and exit shall be located on the shallow side of the first slope change.

809.3 Secondary entries and exits. Where water depth in the deep area of a pool exceeds 5 feet (1524 mm), a means of entry and exit as indicated in Section 809.2 shall be provided in the deep area of the pool.

> **Exception:** Where the required placement of a means of exit from the deep end of a pool would present a potential hazard, handholds shall be provided as an alternative for the means of exit.

809.4 Over 30 feet in width. Pools over 30 feet (9144 mm) in width at the deep area shall have an entry and exit on both sides of the deep area of the pool.

809.5 Pool stairs. The design and construction of stairs into the shallow end and recessed pool stairs shall conform to Sections 809.5.1 through 809.5.3.

809.5.1 Tread dimension and area. Treads shall have a minimum unobstructed horizontal depth of 10 inches (254 mm) and a minimum unobstructed walking surface area of 240 square inches ($0.15\ m^2$).

809.5.2 Riser heights. Risers, other than the top and bottom riser, shall have a uniform height of not greater than 12 inches (305 mm). The top riser height shall be any dimension not exceeding 12 inches (305 mm) for the width of the walking surface. The bottom riser height shall be any dimension not exceeding 12 inches (305 mm). The top and bottom riser heights shall not be required to be equal to each other or equal to the uniform riser height. Riser heights shall be measured at the horizontal centerline of the walking surface area.

809.5.3 Additional steps. In design water depths exceeding 48 inches (1219 mm), additional steps shall not be required.

809.6 Beach and sloping entries. The slope of beach and sloping entries used as a pool entrance shall not exceed 1 unit vertical in 7 units horizontal (14-percent slope).

809.7 Steps and sloping entries. Where steps and benches are used in conjunction with sloping entries, the vertical riser distance shall not exceed 12 inches (305 mm). For steps used in conjunction with sloping entries, the requirements of Section 809.6 shall apply.

809.8 Architectural features. Surfaces of architectural features shall not be required to comply with the 1 unit vertical in 7 units horizontal (14-percent slope) slope limitation.

809.9 Maximum depth. The horizontal surface of underwater seats, benches and swimouts shall be not greater than 20 inches (508 mm) below the design waterline.

SECTION 810
CIRCULATION SYSTEMS

810.1 Turnover rate. The circulation system equipment shall be sized to provide a turnover of the pool water not less than once every 12 hours. The system shall be designed to provide the required turnover rate based on the manufacturer's specified maximum flow rate of the filter, with a clean media condition of the filter.

810.2 Strainer required. Pressure filter systems shall be provided with a strainer located between the pool and the circulation pump.

SECTION 811
SAFETY FEATURES

811.1 Rope and float. In pools where the point of first slope break occurs, a rope and float assembly shall be installed across the width of the pool. The rope assembly shall be located not less than 1 foot (305 mm) and not greater than 2 feet (610 mm) towards the shallow side of the slope break. Rope anchoring devices shall be permanently attached to the pool wall, coping or deck. Rope ends shall attach to the rope anchor devices so that the rope ends can be disconnected from the rope anchor device.

CHAPTER 9

PERMANENT RESIDENTIAL SPAS AND PERMANENT RESIDENTIAL EXERCISE SPAS

User note:

About this chapter: *Chapter 9 regulates permanent residential spas and exercise spas by reference to Chapter 5 and exempting certain sections of that chapter as those regulations are not needed for a residential setting.*

SECTION 901 GENERAL

901.1 Scope. This chapter shall govern the design, installation, construction and repair of permanently installed *residential* spas and exercise spas intended for *residential* use.

901.2 General. Permanent *residential* spas and permanent *residential* exercise spas shall comply with Chapter 5 except that Sections 504.1, 504.1.1, 508.1 and 509 shall not apply. Such spas shall comply with the requirements of Chapter 3.

SECTION 902 SAFETY FEATURES

902.1 Instructions and safety signage. Instructions and safety signage shall comply with the manufacturer's recommendations and the requirements of the local jurisdiction.

CHAPTER 10

PORTABLE RESIDENTIAL SPAS AND PORTABLE RESIDENTIAL EXERCISE SPAS

User note:

About this chapter: *Chapter 10 regulates portable residential spas and exercise spas by requiring compliance with product standards.*

SECTION 1001 GENERAL

1001.1 Scope. This chapter shall govern the installation, alteration and repair of portable *residential* spas and portable exercise spas intended for *residential* use.

1001.2 General. In addition to the requirements of this chapter, portable *residential* spas and portable *residential* exercise spas shall comply with the requirements of Chapter 3.

1001.3 Listing. Equipment and appliances shall be *listed* and *labeled*, and installed as required by the terms of their approval, in accordance with the conditions of the listing, the manufacturer's instructions and this code. Manufacturer's instructions shall be available on the job site at the time of inspection.

1001.4 Certification. Factory-built portable spas and portable exercise spas installed in *residential* applications shall be *listed* and *labeled* in compliance with UL 1563 or CSA C22.2 No. 218.1.

1001.5 Installation. Spa equipment shall be supported to prevent damage from misalignment and settling in accordance with the manufacturer's instructions.

1001.6 Access. Electrical components that require placement or servicing shall be accessible.

1001.7 Instructions and safety signage. Instructions and safety signage shall comply with UL 1563 or CSA C22.2 No. 218.1, the manufacturer's recommendations, and the requirements of the local jurisdiction.

CHAPTER 11
REFERENCED STANDARDS

User note:

About this chapter: *This code contains numerous references to standards promulgated by other organizations that are used to provide requirements for materials and methods of construction. Chapter 11 contains a comprehensive list of all standards that are referenced in this code. These standards, in essence, are part of this code to the extent of the reference to the standard.*

This chapter lists the standards that are referenced in various sections of this document. The standards are listed herein by the promulgating agency of the standard, the standard identification, the effective date and title, and the section or sections of this document that reference the standard. The application of the referenced standards shall be as specified in Section 102.7.

AHRI

Air Conditioning, Heating & Refrigeration Institute
2111 Wilson Boulevard, Suite 500
Arlington, VA 22201

400 (I-P)—2015: Performance Rating of Liquid to Liquid Heat Exchangers
Table 316.2

1160 (I-P)—2014: Performance Rating of Heat Pump Pool Heaters
Table 316.2

ANSI

American National Standards Institute
25 West 43rd Street, 4th Floor
New York, NY 10036

A108/A118/A136.1—2008: Specifications for Installation of Ceramic Tile
Table 307.2.2

Z21.56a/CSA 4.7—2017: Gas Fired Pool Heaters
Table 316.2

APSP

The Association of Pool & Spa Professionals
2111 Eisenhower Avenue, Suite 500
Alexandria, VA 22314

ANSI/APSP/ICC 4—12: American National Standard for Aboveground/Onground Residential Swimming Pools—Includes Addenda A Approved April 4, 2013
702.2.1

ANSI/APSP/ICC 7—13: American National Standard for Suction Entrapment Avoidance in Swimming Pools, Wading Pools, Spas, Hot Tubs, and Catch Basins
310.1

ANSI/APSP/ICC 14—2014: American National Standard for Portable Electric Spa Energy Efficiency
303.2

ANSI/APSP/ICC 15a—2011: American National Standard for Energy Efficiency Residential Inground Swimming Pool and Spas—Includes Addenda A Approved January 9, 2013
303.3

ANSI/APSP/ICC 16—11: American National Standard for Suction Fittings for Use in Swimming Pools, Wading Pools, Spas, and Hot Tubs
202, 311.4.1, 311.4.4, 505.2.1

ASCE/SEI

American Society of Civil Engineers
Structural Engineering Institute
1801 Alexander Bell Drive
Reston, VA 20191-4400

ASCE 24—14: Flood Resistant Design & Construction
304.3

ASME

American Society of Mechanical Engineers
Two Park Avenue
New York, NY 10016-5990

A112.1.2—2012: Air Gaps in Plumbing Systems (For Plumbing Fixtures and Water-connected Receptors)
318.2

B16.15—2013: Cast Alloy Threaded Fittings: Classes 125 and 250
Table 311.4.1

ASTM

ASTM International
100 Barr Harbor, P.O. Box C700
West Conshohocken, PA 19428-2959

A182—15: Standard Specification for Forged or Rolled Alloy and Stainless Steel Pipe Flanges, Forged Fittings, and Valves and Parts for High-temperature Service
Table 311.4.1

A240/A240M—15a: Standard Specification for Chromium and Chromium-nickel Stainless Steel Plate, Sheet and Strip for Pressure Vessels and for General Applications
Table 307.2.2

A312/A312M—15a: Standard Specification for Seamless, Welded, and Heavily Cold Worked Austenitic Stainless Steel Pipes
Table 311.4

A403—15: Standard Specification for Wrought Austenitic Stainless Steel Piping Fittings
Table 311.4.1

B88—14: Standard Specification for Seamless Copper Water Tube
Table 311.4

B447—12a: Specification for Welded Copper Tube
Table 311.4

D1527—99(2005): Specifications for Acrylonitrile Butadiene Styrene (ABS) Plastic Pipe, Schedules 40 and 80
Table 311.4, Table 311.4.1

D1593—13: Standard Specification for Nonrigid Vinyl Chloride Plastic Film and Sheeting
Table 307.2.2

D1785—15: Specification for Poly Vinyl Chloride (PVC) Plastic Pipe, Schedules 40, 80 and 120
Table 311.4

D2241—15: Standard Specification for Poly (Vinyl Chloride) (PVC) Pressure-rated Pipe (SDR Series)

D2464—15: Standard Specification for Threaded Poly (Vinyl Chloride) (PVC) Plastic Pipe Fittings, Schedule 80
Table 311.4.1

D2466—15: Standard Specification for Poly (Vinyl Chloride) (PVC) Plastic Pipe Fittings, Schedule 40
Table 311.4.1

D2467—15: Standard Specification for Poly (Vinyl Chloride) (PVC) Plastic Pipe Fittings, Schedule 80
Table 313.4.1

D2672—14: Standard Specification for Joints for IPS PVC Pipe Using Solvent Cement
Table 311.4

D2846/D2846M—14: Standard Specification for Chlorinated Poly (Vinyl Chloride) (CPVC) Plastic Hot- and Cold-Water Distribution Systems
Table 311.4, Table 311.4.1

F437—15: Standard Specification for Threaded Chlorinated Poly (Vinyl Chloride) (CPVC) Plastic Pipe Fittings, Schedule 80
Table 311.4.1

ASTM—continued

F438—15: Standard Specification for Socket-type Chlorinated Poly (Vinyl Chloride) (CPVC) Plastic Pipe Fittings, Schedule 40
Table 311.4.1

F439—13: Standard Specification for Chlorinated Poly (Vinyl Chloride) (CPVC) Plastic Pipe Fittings, Schedule 80
Table 311.4.1

F1346—91(2010): Standard Performance Specification for Safety Covers and Labeling Requirements for All Covers for Swimming Pools, Spas and Hot Tubs
305.1, 305.4

CPSC

Consumer Product Safety Commission
4330 East-West Highway
Bethesda, MD 20814

16 CFR Part 1207—04: Safety Standard for Swimming Pool Slides
406.10

CSA

CSA Group
8501 East Pleasant Valley Road
Cleveland, OH 44131-5516

B137.2—16: Polyvinylchloride (PVC) Injection-moulded Gasketed Fittings for Pressure Application
Table 311.4, Table 311.4.1

B137.3—16: Rigid Polyvinylchloride (PVC) Pipe and Fitting and Pressure Applications
Table 311.4, Table 311.4.1

B137.6—16: Chlorinated Polyvinylchloride (CPVC) Pipe, Tubing, and Fitting for Hot- and Cold-water Distribution Systems
Table 311.4, Table 311.4.1

C22.2 No. 108—14: Liquid Pumps
313.8

C22.2 No. 218.1—13: Spas, Hot Tubs and Associated Equipment
302.3, 309.1, 310.1, 313.8, Table 316.2, 317.2, 509.1, 1001.4, 1001.7

C22.2 No. 236—15: Heating and Cooling Equipment
Table 316.2

Z21.56a/CSA 4.7—2017: Gas Fired Pool Heaters
Table 316.2

IAPMO

IAPMO
4755 E. Philadelphia Street
Ontario, CA 91761-USA

IAPMO Z124.7—2013: Prefabricated Plastic Spa Shells
Table 307.2.2

ICC

International Code Council, Inc.
500 New Jersey Avenue, NW
6th Floor
Washington, DC 20001

IBC—18: International Building Code®
201.3, 304.2, 306.1, 307.2, 307.4, 307.8, 307.9, 410.1

IECC—18: International Energy Conservation Code®
201.3, 316.4

IFC—18: International Fire Code®
201.3

IFGC—18: International Fuel Gas Code®
201.3, 316.4

IMC—18: International Mechanical Code®
201.3, 316.4

ICC—continued

IPC—18: International Plumbing Code®
201.3, 302.2, 302.5, 302.6, 306.8, 306.8.1, 318.2, 410.1

IRC—18: International Residential Code®
102.7.1, 201.3, 302.1, 302.5, 302.6, 304.2, 306.1, 306.3, 306.8, 306.8.1, 307.2, 307.4, 307.8, 307.9, 316.4, 318.2, 321.2.1, 321.4, 703.1, 802.1, 802.2

ICC 900/SRCC 300—2015: Solar Thermal System Standard
316.6.2

ICC 901/SRCC 100—2015: Solar Thermal Collector Standard
316.6.2

NEMA

National Electrical Manufacturers Association
1300 North 17th Street
Suite 900
Rosslyn, VA 22209

NEMA Z535—2017: ANSI/NEMA Color Chart
409.3

NFPA

National Fire Protection Association
1 Batterymarch Park
Quincy, MA 02169-7471

NFPA 70—2017: National Electrical Code
302.1, 316.4, 321.2.1, 321.4

NSF

NSF International
789 N. Dixboro Road
P.O. Box 130140
Ann Arbor, MI 48105

NSF 14—2015: Plastics Pumping Systems Components and Related Materials
302.3, 311.4

NSF 50—2015: Equipment for Swimming Pools, Spas, Hot Tubs, and Other Recreational Water Facilities
302.3, 309.2, Table 316.2, 508.1

UL

UL LLC
333 Pfingsten Road
Northbrook, IL 60062

372—2007: Automatic Electrical Controls for Household and Similar Use—Part 2: Particular Requirements for Burner Ignition Systems and Components—with revisions through July 2012
506.2.1, 506.2.2

873—2007: Temperature-indicating and Regulating Equipment—with revisions through February 2015
506.2.1, 506.2.2

1004-1—12: Standard for Rotating Electrical Machines General Requirements—with revisions through June 2011
313.8

1081—2008: Standard for Swimming Pool Pumps, Filters and Chlorinators—with revisions through March 2014
313.8

1261—2001: Standard for Electric Water Heaters for Pools and Tubs—with revisions through July 2012
Table 316.2

1563—2009: Standard for Electric Hot Tubs, Spas and Associated Equipment—with revisions through March 2015
302.3, 309.1, 310.1, 313.8, Table 316.2, 317.2, 509.1, 1001.4, 1001.7

1995—2011: Heating and Cooling Equipment—with revisions through July 2015
Table 316.2

2017—2008: General-purpose Signaling Devices and Systems—with revisions through May 2011
305.4

INDEX

U

V

W

EDITORIAL CHANGES – SECOND PRINTING

Page 29, **Section 411.1:** line 5 now reads . . . disabilities in accordance with Section 307.1.4 shall not be

For the complete errata history of this code, please visit: https://www.iccsafe.org/errata-central/

ANSI/APSP/ICC-7 2013

Approved American National Standard ANSI

American National Standard for

Suction Entrapment Avoidance In Swimming Pools, Wading Pools, Spas, Hot Tubs, and Catch Basins

Approved October 8, 2013

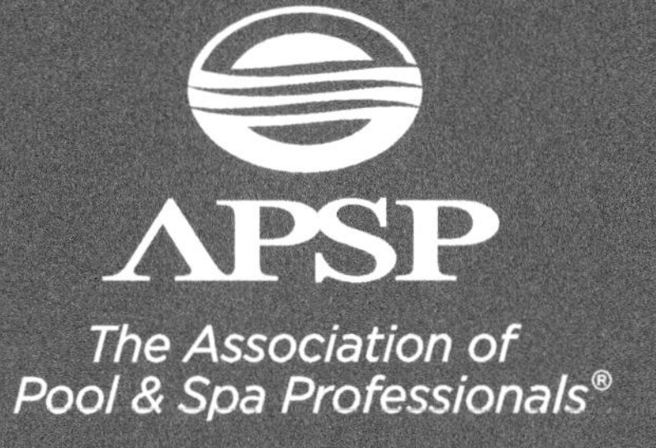

ANSI/APSP/ICC-7 2013

American National Standard for Suction Entrapment Avoidance in Swimming Pools, Wading Pools, Spas, Hot Tubs And Catch Basins

SECRETARIAT:

Association of Pool & Spa Professionals
2111 Eisenhower Avenue
Alexandria, VA 22314
703 838-0083
www.APSP.org

Approved October 8, 2013
American National Standards Institute

American National Standard

Approval of an American National Standard requires verification by ANSI that the requirements for due process, consensus, and other criteria for approval have been met by the standard developer. Consensus is established when, in the judgment of the ANSI Board of Standards Review, substantial agreement has been reached by directly and materially affected interests. Substantial agreement means much more than a simple majority, but not necessarily unanimity.

Consensus requires that all views and objections be considered and that a concerted effort be made toward their resolution. The use of American National Standards is completely voluntary; their existence does not in any respect preclude anyone, whether he has approved this standard or not, from manufacturing, marketing, purchasing, or using products, processes, or procedures not conforming to the standards.

The American National Standards Institute does not develop standards and will in no circumstances give an interpretation of any American National Standard. Moreover, no person shall have the right or authority to issue an interpretation of an American National Standard in the name of the American National Standards Institute. Requests for interpretations should be addressed to the secretariat or sponsor whose name appears on the title page of this standard.

NOTICE: This American National Standard may be revised or withdrawn at any time. The procedures of the American National Standards Institute require that action be taken periodically to reaffirm, revise, or withdraw this standard.

Important Notice about this Document

This voluntary standard has been developed under the published procedures of the American National Standards Institute (ANSI). The ANSI process brings together volunteers representing varied viewpoints and interests to achieve consensus.

APSP does not write the standards. Rather, APSP facilitates a forum for its members, and others interested in pool and spa design and safety, to develop standards through the consensus procedures of the American National Standards Institute (ANSI). While the APSP administers the process and establishes rules to promote fairness in the development of consensus, it does not independently test, evaluate, or verify the accuracy of any information or the soundness of any judgments contained in its codes and standards.

In issuing and making this document available, the APSP is not undertaking to render professional or other services for or on behalf of any person or entity. Nor is the APSP undertaking to perform any duty owed by any person or entity to someone else. The APSP disclaims liability for any personal injury, property, or other damages of any nature whatsoever, whether special, indirect, consequential, or compensatory, directly or indirectly resulting from the publication of, use of, or reliance on this document.

The APSP has no power, nor does it undertake, to police or enforce compliance with the contents of this document. The APSP does not list, certify, test, or inspect products, designs, or installations for compliance with this document. Any certification or other statement of compliance with the requirements of this document shall not be attributable to the APSP. Any certification of products stating compliance with requirements of this document is the sole responsibility of the certifier or maker of the statement. The APSP, its members, and those participating in its activities do not accept any liability resulting from compliance or noncompliance with the provisions given herein, for any restrictions imposed on materials, or for the accuracy and completeness of the text.

Anyone using this document should rely on his or her own independent judgment or, as appropriate, seek the advice of a competent professional in determining the exercise of reasonable care in any given circumstance. It is assumed and intended that spa users will exercise appropriate personal judgment and responsibility and that public spa owners and operators will create and enforce rules of behavior and warnings appropriate for their facility.

Copyright Notice

Foreword

This Foreword is not part of the American National Standard ANSI/APSP/ICC-7 2013. It is included for information only.

The *ANSI/APSP/ICC-7 2013 Standard for Suction Entrapment Avoidance in Swimming Pools, Wading Pools, Spas, Hot Tubs And Catch Basins* was approved by ANSI on October 8, 2013.

The objective of this voluntary standard is to provide recommended minimum guidelines for suction entrapment avoidance in the design, equipment, operation, and installation of new and existing swimming pools, wading pools, spas, hot tubs, and catch basins for builders, installers, pool operators, and service professionals. It is intended to meet the needs for incorporation into national or regional building codes, and also for adoption by state and/or local municipalities as a local code or ordinance. It is understood that for the sake of applicability and enforceability, the style and format of the standard may need adjustment to meet code or ordinance style of the jurisdiction adopting this document.

The design recommendations and construction practices in this standard are based upon sound engineering principles, research, and field experience that, when applied properly, contribute to the delivery and installation of a safe product.

The words "safe" and "safety" are not absolutes. While the goal of this standard is to design and construct a safe, enjoyable product, it is recognized that risk factors cannot, as a practical matter, be reduced to zero in any human activity. This standard does not replace good judgment and personal responsibility. In permitting use of the pool, spa, swim spa or portable spa by others, owners must consider the skill, attitude, training and experience of the expected user.

As with any product, the specific recommendations for installation and use provided by the manufacturer should be carefully observed.

This standard was prepared by the This standard was prepared by the APSP-7 Suction Entrapment Avoidance Standard Writing Committee of the Association of Pool and Spa Professionals (APSP) in accordance with American National Standards Institute (ANSI) Essential Requirements: Due process requirements for American National Standards.

Consensus approval was achieved by a ballot of the balanced APSP Standards Consensus Committee and through an ANSI Public Review process. The ANSI Public Review provided an opportunity for additional input from industry, academia, regulatory agencies, safety experts, state code and health officials, and the public at large.

Suggestions for improvement of this standard should be sent to The Association of Pool and Spa Professionals, 2111 Eisenhower Avenue, Alexandria, VA 22314.

This standard is published in partnership with the International Code Council (ICC). ICC develops and publishes the *International Building Code (IBC)* and *International Residential Code (IRC)*, which are adopted as the basis for the building codes used in most states and jurisdictions within the United States. Additionally, APSP and ICC have collaborated to develop the first comprehensive model swimming pool and spa code, known as the *International Swimming Pool and Spa Code*. This landmark document incorporates and references material from ANSI/APSP standards and ICC's model codes, to create a stand-alone code that is consistent with codes and standards from both organizations.

These codes and standards are the result of a joint effort between ICC and APSP as a service to both the swimming pool and spa community, and building code professionals. It is the hope of both organizations that they will lead to enhanced safety for pool and spa users around the world.

Organizations Represented

Consensus approval in accordance with ANSI procedures was achieved by ballot of the following APSP Standards Consensus Committee. Inclusion in this list does not necessarily imply that the organization concurred with the submittal of the proposed standard to ANSI.

Producers

All American Custom Pools & Spas, Inc . . . John Romano
Custom Pools, Inc . . . . Scott Heusser
Gary Pools, Inc . . . . Leif Zars
Hayward Industries . . . . John O'Hare
HornerXpress South Florida . . . . Bill Kent
Master Spas Inc . . . . Nathan Coelho
Rosebrook Carefree Pools, Inc . . . . John Bently
Trilogy Pools Div. of Viking Pools LCC . . . . . Ted Baudendistel
S.R. Smith, LLC . . . . Bill Svendsen
Van Kirk & Sons, Inc . . . . Don Cesarone

General Interest

American Hotel & Lodging Association . . . . Tony Mendez
American Red Cross . . . . Connie Harvey
Chesapeake Aquatic Consultants, LLC . . . . Frank Goldstein
Con-Serv Associates Inc. . . . . Wally James
Conroe Independent School District, TX . . . Louis Sam Fruia
Don Witte Consultant. . . . . Don Witte
National Environmental Health Association (NEHA) . . . . Florence Higgins
Walt Disney Parks and Resorts. . . . . Michael Beatty
World Waterpark Association . . . . Rick Root
YMCA of the USA. Albert Tursi

Government/User

City of Martinsville, VA. . . . . Kris Bridges
City of Mount Dora, FL. . . . . Tom Allen
Fairfax County, VA . . . . Marc Mordue
Green Oak Charter Township . . . . Wayne Jewell
Illinois Department of Public Health . . . . . . Justin DeWitt
International Code Council. . . . . Lee Clifton
New Jersey Dept. of Community Affairs Division of Codes and Standards . . . . Thomas Pitcherello
North Carolina Building Office Office of State Fire Marshall. . . . . Helen DiPietro
Oregon Public Health Division . . . . Stephen Keifer
Washington State Dept. of Health . . . . Gary Fraser
U.S. Consumer Product Safety Commission Mark Eilbert*
*non-voting

In accordance with American National Standards Institute (ANSI) procedures, this document will be reviewed periodically. The Association of Pool & Spa Professionals welcomes your comments and suggestions, and continues to review all APSP standards, which include:

ANSI/APSP/ICC-1 2013 Standard for Public Swimming Pools
ANSI/APSP-2 1999 Standard for Public Spas
ANSI/APSP-3 1999 Standard for Permanently Installed Residential Spas
ANSI/APSP/ICC-4 a 2013 Standard for Aboveground/Onground Residential Swimming Pools
ANSI/APSP/ICC-5 2011 Standard for Residential Inground Swimming Pools
ANSI/APSP/ICC-6 2013 Standard for Residential Portable Spas and Swim Spas
ANSI/APSP/ICC-7 2013 Standard for Suction Entrapment Avoidance in Recreational Aquatic Vessels
ANSI/APSP/ICC-8 2005 (R2013) Model Barrier Code for Residential Swimming Pools, Spas and Hot Tubs
ANSI/APSP-9 Standard for Aquatic Recreation Facilities (in progress)
ANSI/APSP-11 2009 Standard for Water Quality in Public Pools and Spas
ANSI/APSP/ICC-14 2011 Standard for Portable Electric Spa Energy Efficiency
ANSI/APSP/ICC-15-a 2013 Standard for Residential Pool and Spa Energy Efficiency
ANSI/APSP-16 2011 Suction Fittings for Use in Swimming Pools, Wading Pools, Spas, and Hot Tubs
APSP 2013 Workmanship Standards for Swimming Pools and Spas

Contents

APSP-7 Writing Committee

Zodiac Pool Systems. . . . . Shajee Siddiqui, Chairman
Afras Industries, Inc. . . . . Reza Afshar
Aquatic Development Group . . . . Jim Dunn
Fail Safe Products, LLC . . . . Michael L. Wolfe
Gary Pools . . . . Leif Zars
H_2O Flow Controls . . . . Paul Hackett
Latham International . . . . Michael Tinkler
Newport Controls LLC . . . . Lee West
Pentair Aquatic Systems . . . . Steve Barnes
Professional Pool Solutions, LCC . . . . Maria Bella
Regal Beloit Corporation . . . . Howard Richardson
Stingl Products . . . . David Stingl
Swim, Inc. . . . . Dan Johnson
Vac-Alert . . . . Paul Pennington
Vacless . . . . Hassan Hamza
Walt Disney . . . . Andrea Crabb
Waterway Plastics . . . . Ray Mirzaei

Observers

Custom Molding Products, Inc. . . . . Angelo Pugliese
IAPMO R & T Lab . . . . Tony Zhou
International Code Council (ICC) . . . . Lee Clifton
ICC Alternate . . . . Maribel Campos
Intex Recreation Corp. . . . . Matthew (Chip) Whalen
Master Spas Inc. . . . . Nathan Coelho
Tropical Pools Inc. . . . . Adam Alstott
U.S. Consumer Product Safety Commission Mark Eilbert*
Vak Pak. . . . . Alex Fletcher
Watkins Manufacturing . . . . Mike McAgue

* Non-voting

Also Contributing

Consultant . . . . Ray Cronise
Hayward Pool Products . . . . Robert Rung

APSP Staff

Bernice Crenshaw, Director, Technical and Standards
Carvin DiGiovanni, Senior Director, Technical and Standards

American National Standard for Suction Entrapment Avoidance in Swimming Pools, Wading Pools, Spas, Hot Tubs And Catch Basins

1 Scope

1.1 General. This standard covers design and performance criteria for circulation systems including components, devices, and related technology installed to protect against entrapment hazards in residential and public swimming pools, wading pools, inground spas, infinity edge basins, (infinity edge type pools) and catch Pools, and Aquatic Recreation Facilities.

1.1.1 Portable Factory Built Electric Spas/Hot Tubs. Suction entrapment avoidance guidelines for portable electric spas/hot tubs are not covered by this standard they are covered by UL 1563, Electric Spas, equipment Assemblies, and Associated Equipment.[6]

1.1.2 This standard applies to new and, when retrofitting, existing installations.

1.1.3 DANGER! SUCTION ENTRAPMENT HAZARD: To avoid serious injury or death, the pool or spa shall be closed to bathers if any suction outlet cover/grate is missing, broken, or incorrectly installed. There is no backup for a missing, damaged or incorrectly installed suction outlet cover/grate. See Appendix C.

1.2 Alternative methods. The provisions of this standard are not intended to prevent the use of any alternative material, system, or method of construction, provided any such alternative meets the intent and requirements of this standard, follows manufacturer's product specific instructions and is approved by the authority having jurisdiction.

2 Normative references

The following standards contain provisions that, through reference in this text, constitute provisions of this standard.

ANSI/APSP-16 2011, Suction fittings for swimming and wading pools, spas, hot tubs and whirlpool bathtub appliances[1]

ANSI/ASME A112.19.17-2010, Manufactured safety vacuum release systems (SVRS) for residential and commercial swimming pool, spa, hot tub and wading pool suction systems[2]

ASTM F 2387-12, Standard specification for manufactured safety vacuum release systems, swimming pools, spas and hot tubs[3]

IAPMO SPS-4 2009, Special use suction fittings for swimming pools, spas and hot tubs (for suction side automatic swimming pool cleaners)[4]

NFPA 70-2011, National Electrical Code, Article 680, Swimming pools, fountains, and similar installations[5]

UL 1563 2009, Electric Spas, Equipment Assemblies, and Associated Equipment.[6]

3 Definitions

alternative method: A substitute way of achieving the same goal or purpose.

anti-entrapment cover: See CERTIFIED SUCTION OUTLET COVER/GRATE.

anti-vortex cover: An outlet cover designed to prevent air entrainment from the surface of the water. This term is no longer used to describe CERTIFIED SUCTION OUTLET COVER/GRATE.

approved safety outlet cover: See CERTIFIED SUCTION OUTLET COVER/GRATE.

automatic pump shut-off system (APSS): A pump motor control or other device capable of turning off, stopping, or otherwise incapacitating a pump(s) in response to a condition (i.e., high vacuum, low flow, low current, etc.) that would indicate that a suction entrapment event has occurred.

branch piping: All pipe and fittings, including the "run" of the junction tee, located between multiple suction outlets fitting (see *Figures 1* and *9–14*).

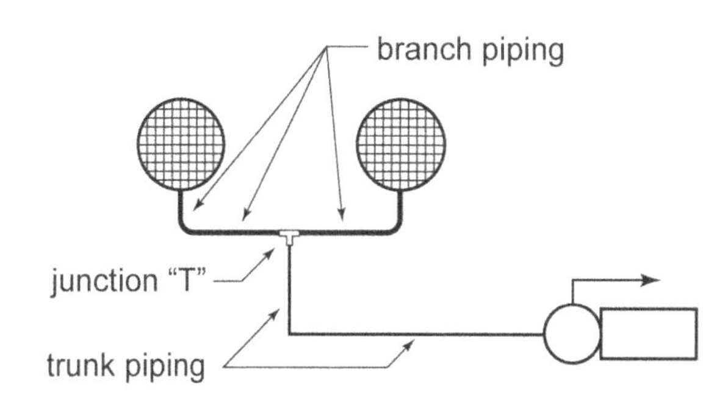

Figure 1. Branch piping

1. Association of Pool & Spa Professionals (APSP) , 2111 Eisenhower Avenue, Alexandria, VA 22314 (703) 838-0083, www.APSP.org.
2. American Society of Mechanical Engineers (ASME), 3 Park Avenue, 20th Floor, New York, NY 10016, (212) 591-8562, www.asme.org
3. ASTM International, 100 Barr Harbor Drive, West Conshohocken, PA 19428, (610) 832-9500, www.astm.org
4. International Association of Plumbing and Mechanical Officials (IAPMO), 5001 E. Philadelphia St., Ontario, CA 91761. (909) 472-4100, www.iapmo.org
5. National Fire Protection Association (NFPA), 1 Batterymarch Park, Quincy, MA 02169-7471, (617) 770-3000, www.nfpa.org
6. Underwriters Laboratories Inc. (UL), 333 Pfingston Road, Northbrook, IL 60062-2096. (847) 272-8800, www.ul.com

catch pool: The pool at the discharge of a waterslide or similar aquatic facility or a body of water supplied by gravity overflow from another pool or vessel.

CAUTION: Indicates a potentially hazardous situation that, if not avoided, could result in minor or moderate injury. It may also be used to alert against unsafe practices.

certified: The published certification by an ILAC approved laboratory that a device, system, or alternate method has been tested and certified to be in conformance with the full intent of a standard.

certified automatic pump shut-off system APSS: An automatic pump shut-off system tested and certified in accordance with Section 4.3.2 of this standard which requires compliance with ANSI/ASME A112.19.17 or ASTM F2387.

certified safety cover: See CERTIFIED SUCTION OUTLET COVER/GRATE.

certified safety outlet cover: See CERTIFIED SUCTION OUTLET COVER/GRATE.

certified suction outlet cover/grate: A manufactured suction outlet or field fabricated outlet that has been Certified in accordance with Section 4.3.1 of this standard which requires compliance with ANSI/APSP-16[7].

certified SVRS: A manufactured safety vacuum release system tested and certified in accordance with Section 4.3.2 of this standard which requires compliance with ANSI/ASME A112.19.17 or ASTM F 2387.

check valve: A mechanical device in a pipe that permits the flow of water in one direction only.

closed pool: A pool in which access to bathers is prohibited. This may be accomplished by locking gates and doors, by posting notices, conspicuously placed "Barricade" tape. Pool circulation systems may be in operation when closed.

DANGER: Indicates an imminently hazardous situation that, if not avoided, will result in death or serious injury.

debris removal system: A system comprised of a large opening suction outlet, large diameter pipe and a debris collection basket, typically located in the deck or the pump basket. Because of unique challenges passing debris through the suction outlet cover/grate and suction piping, these systems are designed specifically for debris removal and are commonly sold as kits with detailed installation requirements to address suction safety.

drain: See MAIN DRAIN.

effluent: The outflow of water from a filter, pump, or pool

engineer: A Licensed Professional Engineer (P.E.).

equalizer line: 1. A pipe with a Certified suction outlet cover/grate located below the waterline and connected to the body of a skimmer to prevent air from being drawn into the pump if the water level drops below the skimmer weir. 2. A pipe connecting two bodies of water with Certified Suction Outlet Fitting Assemblies to equalize water levels.

feet of head: The measure of resistance in a hydraulic system based on the equivalent to the height of a column of water that causes the same resistance (100 feet of head = 43 pounds per square inch).

field built sump: A sump built below or behind the suction outlet cover/grate of a design specified by the cover/grate manufacturer to control flow distribution through the open area of the cover/grate. Field built sumps may be formed, cut or carved out of the gunite or concrete material of the pool structure. They are to be constructed so as to accommodate suitable fastening means to attach the cover/grate. They must conform to the requirements of APSP/ANSI-16.

field fabricated outlet(s): These are site specific unblockable suction outlet fitting assemblies defined per ANSI/APSP-16 as being intended as but not limited to a single suction outlet and are limited to 1.5 ft/sec *(0.46 m/s)* of flow velocity through the open area of the cover/grate unless rated at a lower flow rate by the Registered Design Professional. They are to be of such a size that the 18 in. × 23 in. × 4 in. corner radii *(457 mm × 584 mm × 102 mm)* blocking element will not cause a differential pressure that could cause body entrapment.

flow rate: The quantity of water flowing through a pipe within a specified time, such as the number of gallons flowing past a point in one minute; abbreviated as GPM or *liters/minute, Lpm* (1 GPM = *3.7854 L/min*).

flow rating: The maximum allowable flow rate through a cover/grate.

GPM: Abbreviations for gallons per minute.

gravity drainage: See VENTED RESERVOIR.

gutter: Overflow trough at the perimeter wall of a pool or at the bottom of a vanishing edge wall of a pool that is a component of the circulation system or flows to waste.

hydrostatic relief valve: A valve to allow rising groundwater to enter an empty pool to prevent flotation.

incorrectly installed: not installed in strict conformance with manufacturers product specific instructions.

infinity edge basin: a basin designed to receive the water that flows over the "infinity edge" of a pool and/or spa during the circulation pump cycle and/or feature pump operational time.

influent: The water entering a filter or other device

inlet: See RETURN INLET.

junction tee: A tee between suction outlet fittings, which joins their flow into a trunk line to the pump.
NOTE: This usage is different from the standard usage in the piping industry. The trunk line is connected to the single branch of the tee fitting and the two branch lines from the outlets are connected to the run of the tee fitting.

main drain: An obsolete term for suction outlet, which is a fitting, fitting assembly, cover/grate, and related components that provide a localized low-pressure area for the transfer of water from a swimming pool, wading pool, spa, or hot tub.

manifold: A branch pipe arrangement that connects several influent pipes into one chamber or pump, or one chamber into several effluent pipes.

7. The Consumer Product Safety Commission has voted unanimously to approve ANSI/APSP-16 2011 as the successor standard to the ANSI/ASME A112.19.8 suction outlet cover standard mandated by the Virginia Graeme Baker Pool and Spa Safety Act. The Commission determined that the new standard, ANSI/APSP-16 2011, was in the public interest, and incorporated this standard into its regulations. This means that, effective September 6, 2011, suction outlet covers manufactured, distributed, or entered into commerce in the United States must conform to the requirements of ANSI/APSP-16 2011.

maximum system flow rate: For purposes of this suction entrapment avoidance standard, maximum system flow rate is defined as the maximum potential flow when all available system flow is directed through the submerged suction outlet(s). See Section 4.4.9 for specific procedures required to determine the system specific, maximum system flow rate.

operating point: The condition at which the pump will operate. It is the intersection of the pump curve and system curve.

overflow system: An outlet with flow across a fixed or movable weir and where there is a free surface interface with atmosphere.

P.E.: See ENGINEER.

parallel: A piping arrangement allowing flow through multiple paths.

pump: A mechanical device, usually powered by an electric motor that causes hydraulic flow and pressure for the purpose of filtration, heating, and circulation of pool and spa water. Typically a centrifugal pump is used for pools, spas, and hot tubs.

pump curve, pump performance curve: A graph that represents the pressure rise of a pump plotted against flow rate. See SYSTEM CURVE AND OPERATING POINT.

Registered Design Professional: an individual who is registered or licensed to practice their respective design profession as defined by the statutory requirements of the professional registration laws of the state or jurisdiction in which the project is to be constructed.

retrofit: The act of adding a component or accessory to the pool and spa that was not part of the original installation—for example, replacing a non-certified suction outlet cover/grate with one that is Certified. See also Section 6.5.

return inlet: The aperture or fitting through which the water under pressure returns into the pool or spa.

safety drain cover: See CERTIFIED SUCTION OUTLET s/GRATE.

safety vacuum release system (SVRS): A system capable of providing vacuum release at a suction outlet in case of a high vacuum occurrence due to a suction outlet flow blockage. Methods may include, but are not necessarily limited to, venting the suction line to atmosphere and/or turning off the circulation pump, or reversing the circulation flow.

secured control system: Any means that reasonably prevents unauthorized access to pump and valve control systems by persons who could make adjustments resulting in flow rates above which the system has been stamped and sealed in accordance with this standard by the Registered Design Professional responsible for this system.

NOTE: Secured control systems include, but are not limited to; equipment rooms not accessible to unqualified persons, control systems that are protected by passwords not available to unqualified persons, and valves with adjustment handles locked.

single outlet, alternative suction systems: A single Certified suction outlet cover/grate and an alternative suction system, including a venturi-driven system, turbine driven system, or any other mechanical means of circulating water without the use of a centrifugal pump.

skimmer: A device installed in the wall of a body of water that permits the removal of floating debris and surface water.

static lines: Piping that connects two bodies of water to maintain equal levels (example—a static line from a collector tank to a pool so that the auto-fill device in the collection tank can be adjusted to maintain the proper water level in the pool.)

suction: The flow of fluid into a partial vacuum or region of lower pressure. The gradient between this region and the ambient pressure will propel matter towards the low-pressure area.

suction-limiting gravity flow systems: See VENTED RESERVOIR.

suction-limiting system: A safety vacuum release system, vent system, gravity drainage/flow system, vented reservoir, automatic pump shut-off system, properly spaced multiple suction outlets, or other methods capable of limiting the duration of a high-vacuum occurrence and/or the magnitude of the vacuum at a suction outlet cover/grate in case of suction flow blockage.

suction outlet: Indicates a fitting, fitting assembly, cover/grate, sump, and related components that provide a localized low-pressure area for the transfer of water from a swimming pool, wading pool, spa, or hot tub. See also CERTIFIED SUCTION OUTLET COVER/GRATE.

suction system piping: All piping on the suction side of the system between the pool and the pump.

sump: The vessel between the suction outlet cover/grate and suction outlet piping. This may be manufactured or field built.

sumps in series: An arrangement of outlets such that effluent of one sump is influent to another sump. It is commonly used in piping submerged suction outlet(s) to skimmer body(ies).

swim jet system: Combination fitting or fittings that incorporate(s) a suction outlet and inlet designed to move a large volume of water at high velocity in a single direction.

system curve: A graph that shows the pressure difference required to induce flow through the entire piping system. It is plotted with head pressure on the vertical axis of the chart and flow rate on the horizontal axis of the chart (see *Figure 2*).

tee: A fitting in the shape of a "T" used to connect branch pipes. The trunk pipe is perpendicular to the two branch pipes.

testing: For the purposes of this standard, "testing" means the physical activity of performing an evaluation in accordance with the procedures and protocols defined by this standard and/or a referenced standard.

total dynamic head (TDH): The sum of the difference in elevation between the source and destination and the friction losses in a piping system. It has units of pressure (such as psi) but is commonly given in feet of head. Since friction losses depend on flow rate, TDH must be specified for a particular flow rate.

trunk line: piping from a junction tee to a suction source, such as a pump or vented reservoir.

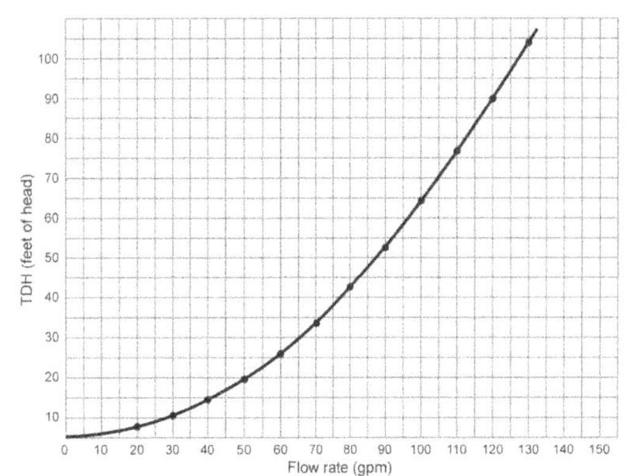

Figure 2. A system curve

unblockable: A suction outlet defined as all components, including the sump and/or body, cover/grate, and hardware such that its perforated (open) area cannot be shadowed by the area of the 18 × 23 in. *(457 × 584 mm)* Body Blocking Element of ANSI/APSP-16, and that the rated flow through the remaining open area cannot create a suction force in excess of the removal force values in *Table 1* of that standard. All suction outlet covers, manufactured or field-fabricated, are to be certified as meeting the applicable requirements of the ANSI/APSP-16 (see *Figures 3a* and *3b*).

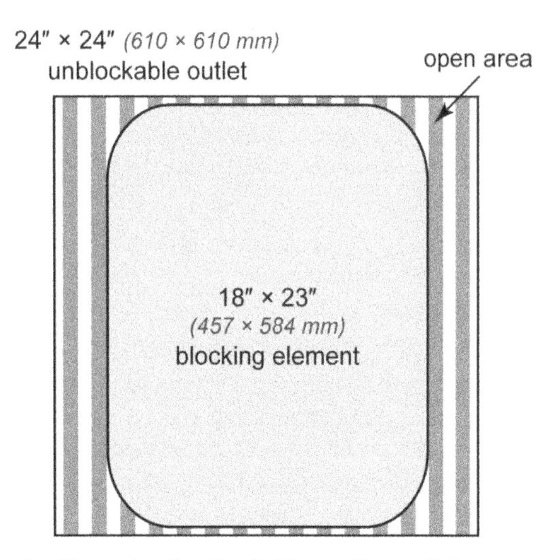

Figure 3a. Example of unblockable suction outlet

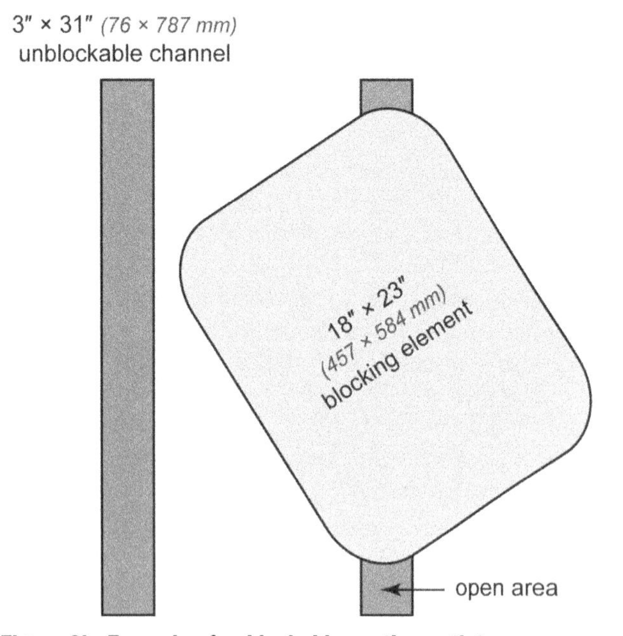

Figure 3b. Example of unblockable suction outlet

vacuum: A condition in which the pressure inside an outlet or suction pipe is lower than pool pressure.

vanishing edge: A design feature incorporated into a pool wall wherein the water flows over the wall (edge) into a catch gutter or catch pool, creating the illusion that the water vanishes.

vent: A vent to the atmosphere that connects to the suction pipe between the pool and the pump. When a high vacuum event occurs, air from the vent pipe replaces the water in the suction pipe thereby breaking the vacuum.

vented reservoir: A receptacle or container incorporated as part of a circulation system that is vented to atmosphere and receives water from the pool/spa or water feature by force of gravity, from which the pump draws its water supply. Systems including vented reservoirs are commonly referred to as *gravity flow systems, gravity feed systems,* or *gravity drainage systems.* Vented reservoirs include, but are not limited, to the following: catch pools, surge tanks, collector tanks, skimmers open to the atmosphere, atmospheric vent, gutters, overflow gutters, or perimeter gutter systems.

wading pool: A separate pool designed for use by small children with an independent circulation system and a maximum water depth of 18 in. *(457 mm)*.

wall vacuum fitting: A fitting in the wall of a pool intended to provide a point of connection of suction for suction side cleaners.

WARNING: Indicates a potentially hazardous situation that, if not avoided, could result in death or serious injury.

4 General requirements for suction entrapment avoidance systems and components

4.1 DANGER! To avoid serious injury or death, the pool or spa shall be closed to bathers if any suction outlet cover/grate is missing, broken, or incorrectly installed. There is no backup for a missing, damaged or incorrectly installed suction outlet cover/grate.

4.2 Codes. Pools and spas covered by this standard shall be constructed and operated to comply with all applicable codes governing safety and environmental regulations.

4.2.1 Electrical components. All associated electrical components installed in and/or adjacent to the circulation system shall comply with the requirements of the *National Electrical Code,* Article 680, Swimming pools, fountains, and similar installations, or the applicable revision and any state or local codes.

4.3 Certifications

4.3.1 Suction outlet certification

4.3.1.1 Manufactured suction outlet fitting assembly(ies). When used, fully submerged suction outlet fitting assembly(ies) including cover/grate and associated fittings, fasteners and components shall be tested and certified by a third-party test lab accredited by the International Laboratory Accreditation Cooperation (ILAC) to test and certify products as conforming to ANSI/APSP-16.

4.3.1.2 Field fabricated suction outlet(s). When used, field fabricated suction outlet cover/grate, sump, fasteners and assemblies shall be Certified by a Registered Design Professional as conforming to ANSI/APSP-16.

4.3.2 Manufactured Safety Vacuum Release Systems (SVRS) and Automatic Pump Shut-off Systems (APSS). When used, SVRS and APSS devices shall be tested and certified by a third-party test lab accredited by the International Laboratory Accreditation Cooperation (ILAC) to test and certify products as conforming to ASME/ANSI A112.19.17, ASTM F 2387 or any successor standards recognized by the U.S. Consumer Product Safety Commission (CPSC).

NOTE: As of the publication date of this standard, automatic pump shut-off systems do not have a performance standard to which they can be certified, as a result the U.S. Consumer Product Safety Commission (CPSC) states APSS are to be tested and certified in accordance with one of the SVRS standards.

NOTICE: Operating conditions. Systems are tested for operation, in accordance with current standards, at room temperature. For substantially varying environmental conditions, including freezing, heat, salt spray, and humidity, confirm suitability with the SVRS manufacturer prior to installation and use.

CAUTION: Incompatible configurations. Some suction vacuum release systems may be incompatible with certain system configurations. The designer or installer shall confirm suitability with the SVRS manufacturer prior to installation and use. Incompatible configurations may include check valves; two or more suction outlets, hydrostatic relief valves, skimmers, solar systems, elevated or submerged pump suction, multilevel bodies of water, and water features.

4.4 Performance requirements for suction outlets and suction-limiting systems

NOTE: Suction-limiting systems protect against body entrapment but are not considered "backup" systems as there is no known suction-limiting system that will completely protect against the remaining four (evisceration, limb, hair, mechanical) of the five known hazards and presenting suction-limiting systems as "backup" systems would promote a false sense of security among the users of these devices/systems.

4.4.1 Submerged suction outlets are optional. Fully submerged suction outlets (main drains) are not required in pools and spas. Surface skimming or overflow systems shall be permitted to provide 100 percent of the flow.

4.4.2 Field built sumps. Field built sumps shall be built in accordance with the suction outlet fitting assembly manufacturer's instructions or as may be site specific designed by a Registered Design Professional.

Figure 4. Single unblockable channel outlet to single pump.

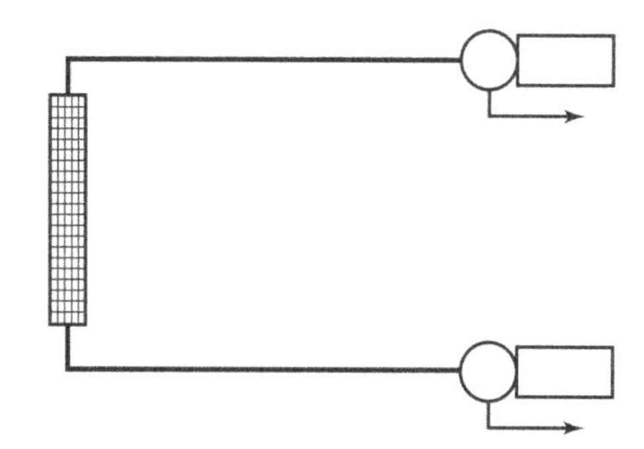

Figure 5. Single unblockable channel outlet to two pumps.

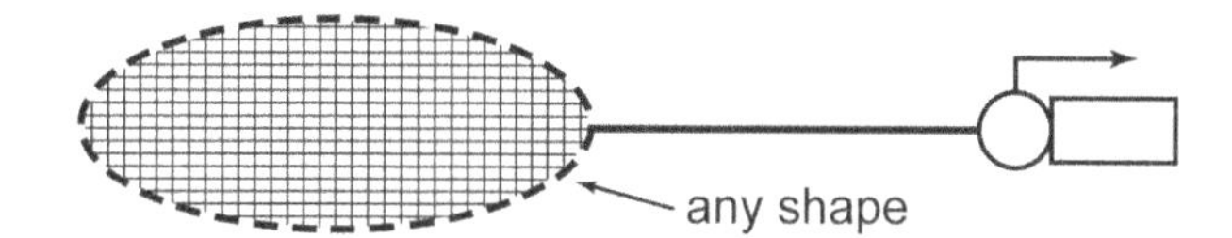

Figure 6. large unblockable outlet of any shape to single pump.

4.4.3 Unblockable outlets–single. A fully submerged unblockable outlet shall have a flow rating equal to or greater than the maximum system flow as determined in accordance with Section 4.4.9 (see *Figures 4, 5,* and *6*).

4.4.4 Unblockable outlets–multiple. Fully submerged unblockable outlets shall have a combined flow rating equal to or greater than the maximum system flow as verified in accordance with Section 4.4.9.

4.4.5 Blockable outlets–dual. When the secondary source of water for a blockable outlet is another submerged suction outlet assembly, each shall have a flow rating equal to or greater than the maximum system flow as determined in accordance with Section 4.4.9.

4.4.6 Blockable outlets–three, or more. When the secondary source of water for a blockable outlet is two or more submerged suction outlet assemblies, the flow rating of the set shall be determined by combining the flow rating of all outlets, minus the flow rating of one. The combined flow rating of the set shall be equal to or greater than the maximum system flow as determined in accordance with Section 4.4.9. If the flow ratings of all outlets are not equal, subtract the flow rating of the outlet with the highest flow rating.

4.4.7 Blockable outlets–multiple separation. For new construction see Section 5.3. For existing pools and spas see Section 6.9.

4.4.9 Maximum system flow rate. The maximum system flow rate shall be determined according to control system type where facilities with unsecured control systems use the options in 4.4.9.1 and facilities with secured control systems use the options in 4.4.9.2:

4.4.9.1 Maximum system flow rate–unsecured control systems. The maximum system flow rate is the pump's flow rate at the highest user selectable speed while the system is configured to operate at the lowest achievable system TDH when all flow is from the submerged suction system (skimmers off), the filter(s) is clean (when included), and all pressure-side valves are in the open (maximum flow) position. This operating point is determined by one of the following:

- Measuring with flow meter accurate to ±10% and installed according to the manufacturers specification, or
- Computing using complete system TDH calculations and then looking up the flow rate using the manufacturers certified pump curve, or
- Measuring system TDH at the pump's drain plugs and then looking up the flow rate using the manufacturer's certified pump curve.

4.4.9.2 Maximum system flow rate–secured control systems. The maximum system flow rate is the flow rate for the pump at its highest operating speed with the lowest operating system resistance as defined by the Registered Design Professional. It applies to new or replacement pumps. Measurements shall be made with a properly sized flow meter accurate to ±10% and installed according to manufacturer's instructions.

NOTE: The flow meter must be installed in accordance with the manufacturer's specific instructions. Careful consideration must be given to pipe diameter and the required straight pipe distances between the flow meter and other fittings such as, but not limited to, elbows, tees, valves etc.
No offset or estimation is to be allowed for flow meters that are not installed in accordance with the manufacturer's instructions. The manufacturer's claims must be NIST[8] traceable and verified by a third party.

4.5 Skimmers. Skimmers shall be vented to the atmosphere through openings in the lid, or through a separate vent pipe (see *Figure 7*).

4.5.1 Skimmer equalizer lines. Skimmer equalizer lines shall not be used on new construction. Existing equalizer(s) shall comply with all submerged suction outlet requirements of this standard (see *Figure 8*).

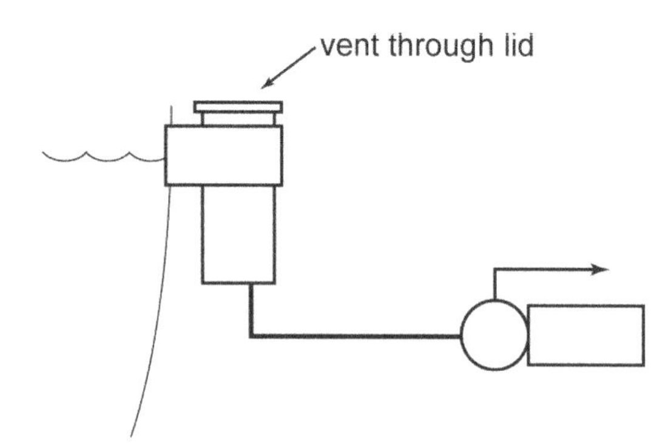

Figure 7. Skimmer, vent through lid

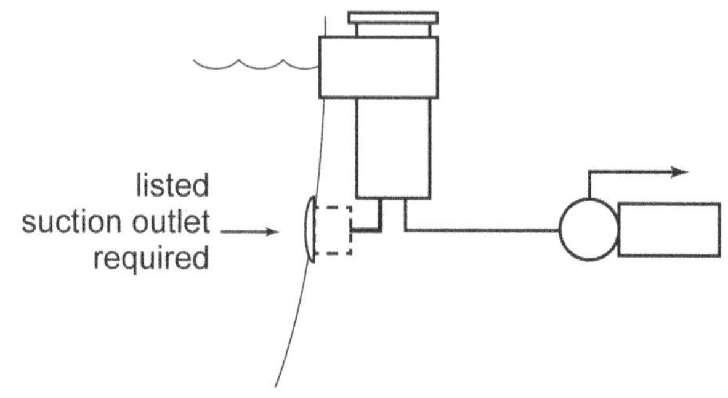

Figure 8. Skimmer with equalizer

4.6 Wall vacuum fitting(s). When used, vacuum cleaner fitting(s) shall be located in an accessible position(s) no greater than 12 inches *(305 mm)* below the water level and the self closing, self latching fitting shall comply with IAPMO SPS 4. In addition, the vacuum piping shall be equipped with a valve to remain in the closed position when not in use.

NOTICE: SPS-4 requires tools to remove, but due to incompatible components, there have been multiple cases of removal upon each usage, sometimes resulting in loss of components, and the essential safety feature. Make sure that the attachment of a vacuum hose in normal usage never leads to removal of the self-closing self-latching feature.

5 New construction

5.1 General. Methods to avoid entrapment in circulation systems, swim jet systems, alternative suction systems, and debris removal systems are shown in Sections 5.2 through 5.5.

5.2 Submerged suction outlets are optional. See Section 4.4.1.

5.2.1 Wading Pools. Due to the unique hazard presented by submerged suction outlets in wading pools, submerged suction outlets are prohibited in wading pools in all areas accessible to the bather.

8. National Institute of Standards and Technology

5.3 Submerged suction outlets. When used, fully submerged suction outlet fitting assemblies and systems shall be certified in accordance with Section 4.3.1. Dual or multiple outlets piped in a single suction system through a common suction line to a pump(s) shall not be capable of being isolated by valves.

5.3.1 Blockable outlets–dual separation. Dual outlets shall be separated by a minimum of 3 feet *(914 mm)* measured from center to center of the suction outlet cover/grate (see *Figures 9, 10,* and *11)* or located on two (2) different planes, i.e., one (1) on the bottom and one (1) on the vertical wall, or one (1) each on two (2) separate vertical walls. (See *Figures 12* and *15*). Suction outlets shall not be installed in seating areas.

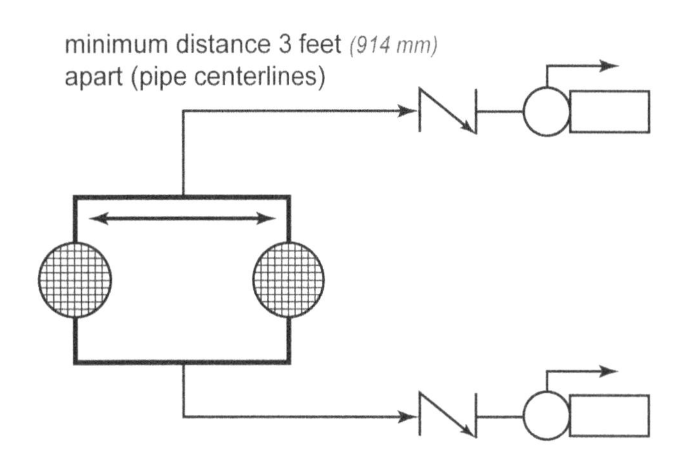

Figure 11. Dual parallel outlets to two pumps

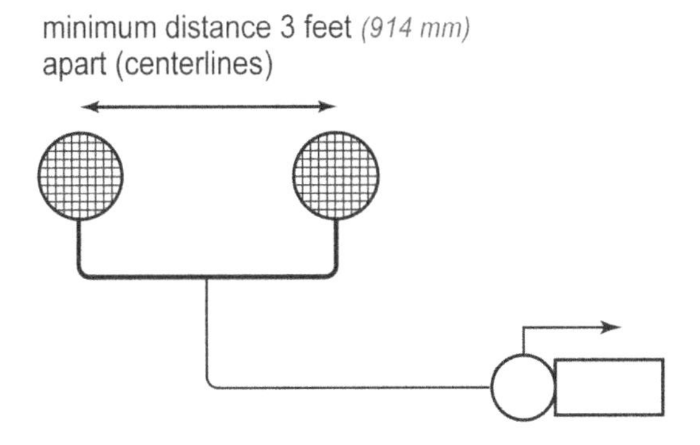

Figure 9. Dual outlets in parallel to one pump

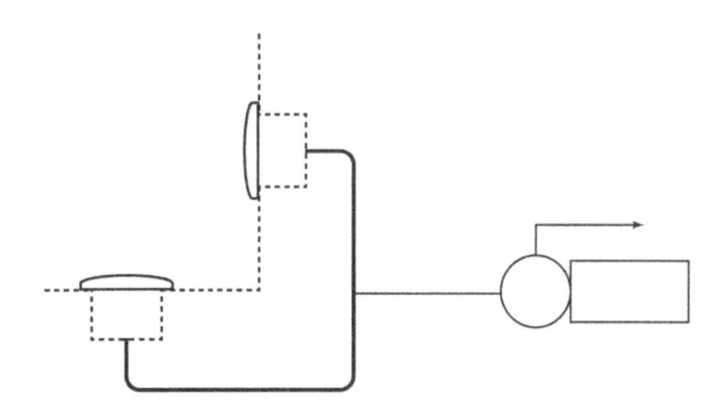

Figure 12. Dual outlets on different planes

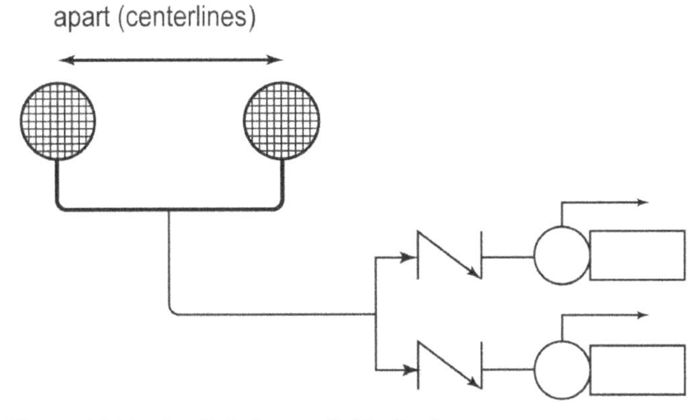

Figure 10. Dual outlets in parallel to dual pumps

5.3.2 Multiple Blockable Separation. Three or more submerged outlets are subject to the separation requirement of 5.3.1 only on the most widely spaced of the group. (See *Figure 13* or *14*.)

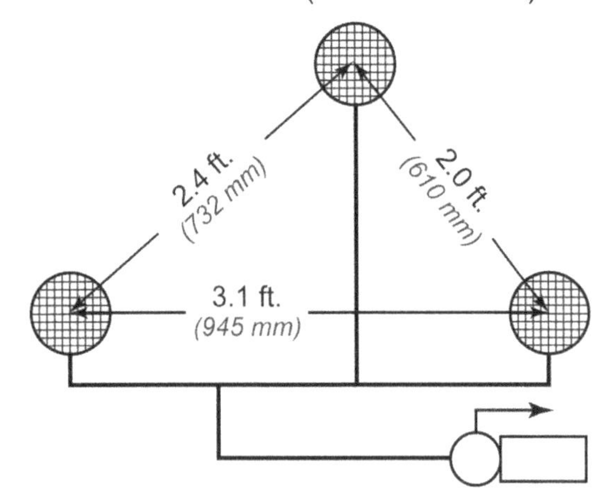

Figure 13. Three or more outlets to (a) single pump(s)

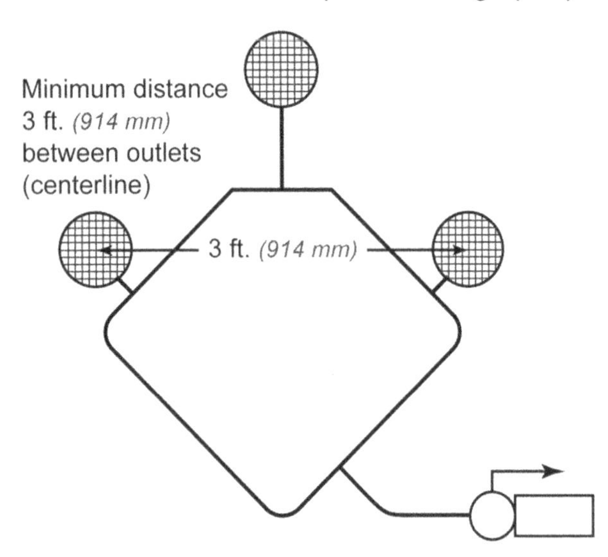

Figure 14. Three or more outlets in parallel, looped piping

5.4 Outlet sumps in series. Two manufactured sumps or field-fabricated sumps, with certified suction outlet covers/grates, piped in series, are typically intended for debris removal. Between the debris suction outlet and the pump, there shall be one of the options certified (see *Figure 15*). The manufacturer of such debris removal systems shall test and approve for the purpose at least one of these.

5.4.1 One (1) additional suction outlet (not in series) with Certified suction outlet cover/grate located a minimum of 18 inches *(457 mm)* from the suction outlet fitting assembly in the suction line to the pump(s); or

5.4.2 Engineered suction-limiting vent system; or

5.4.3 Certified manufactured SVRS or APSS.

5.5 Other means. See Section 1.2.

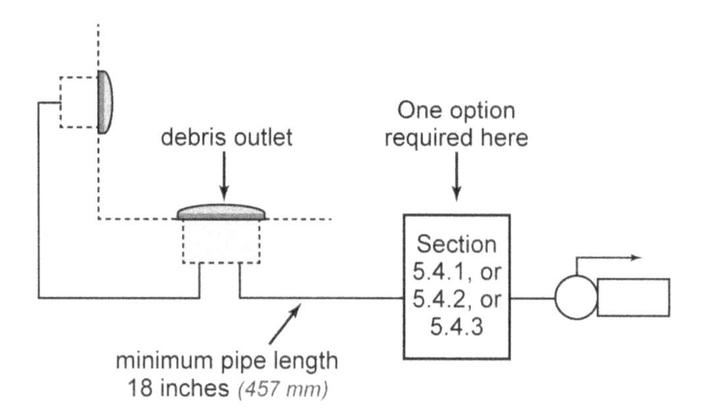

Figure 15. Sumps in series

6 Existing pools and spas

6.1 DANGER! To avoid serious injury or death, the pool or spa shall be closed to bathers if any suction outlet cover/grate is missing, broken, or incorrectly installed. There is no backup for a missing, damaged or incorrectly installed suction outlet cover/grate. See Appendix C.

6.2 Certified Suction Outlet Covers/Grates. When used, fully submerged suction outlet fitting assemblies, cover/grate and associated fitting, fasteners and assemblies shall be certified in accordance with Section 4.3.1, not exceed their installed life in years as indicated by the Certified manufactured or field fabricated outlet documentation, and located in accordance with Section 5.3.

6.3 Wading Pools. Due to the unique hazard presented by submerged suction outlets in wading pools, it is recommended that, whenever possible, the submerged suction outlet(s) be permanently disabled, or converted to a return fitting(s) in accordance with 6.6.1, provided the system piping and skimmer(s)/overflow gutters are capable of handling the required full system flow

6.4 Evaluation for compliance. All suction outlets, suction entrapment avoidance systems, and related components shall be evaluated and brought into compliance by a person who is licensed or qualified by the authority having jurisdiction.

6.5 Retrofitting suction system piping or outlets. When retrofitting, the retrofit installations shall be permitted to utilize a portion of the existing facility and add or replace other elements. The retrofit shall be in accordance with applicable sections of this standard.

6.6 Existing installations–single blockable outlets. The options of Sections 6.6.1 through 6.6.3 shall be permitted.

6.6.1 Convert suction outlet to return inlet by changing the piping and installing an appropriate floor (or wall) inlet(s), designed and/or approved by the manufacturer for that purpose, provided the system piping and skimmer(s) shall be capable of handling the full system flow, in accordance with Section 4.4.9.

6.6.2 Permanently disable the single outlet, provided the system piping and skimmer(s) shall be capable of handling the minimum system flow in accordance with Section 4.4.9. Methods shall include, but not be limited to: permanently plug or cap the suction outlet, or permanently disconnect it from pool pump suction.

6.6.3 When retrofitting existing installations with a single blockable suction outlet, the system shall be retrofitted with either a Certified unblockable suction outlet or a Certified blockable suction outlet cover/grate and at least one of the following:

- Manufactured SVRS or APSS in accordance with Section 4.3.2; or
- Suction-limiting vent system; or
- Gravity drainage/flow system; or
- One or more additional certified suction outlet cover/grate certified in accordance with Section 4.3.1 and located in accordance with Section 5.3.
- Other system approved by the CPSC.

6.7 Existing skimmer equalizer lines. Existing equalizer lines, when used, shall be retrofitted to comply with Section 4.5.

6.8 Existing single blockable outlet piped through skimmer. A single blockable suction outlet piped through a skimmer shall comply with Section 6.6.

6.9 Existing installation–two or more outlets flowing through a common line to pump(s). When retrofitting existing installations, each submerged suction outlet shall comply with Sections 4.3.1 and 4.4.9.

6.9.1 Multiple outlet separation. When existing blockable outlets do not comply with Section 5.3, including applicable subsections 5.3.1 and 5.3.2; the system shall be considered a single blockable outlet system, requiring compliance with Section 6.6.

6.10 Winterization. CAUTION shall be exercised when pools are reopened. All winterizing plugs shall be removed, suction outlet cover/grates shall be secured in place in accordance with manufacturer's instructions, and any safety systems shall be functioning in accordance with manufacturer's instructions.

Appendix A: Symbols

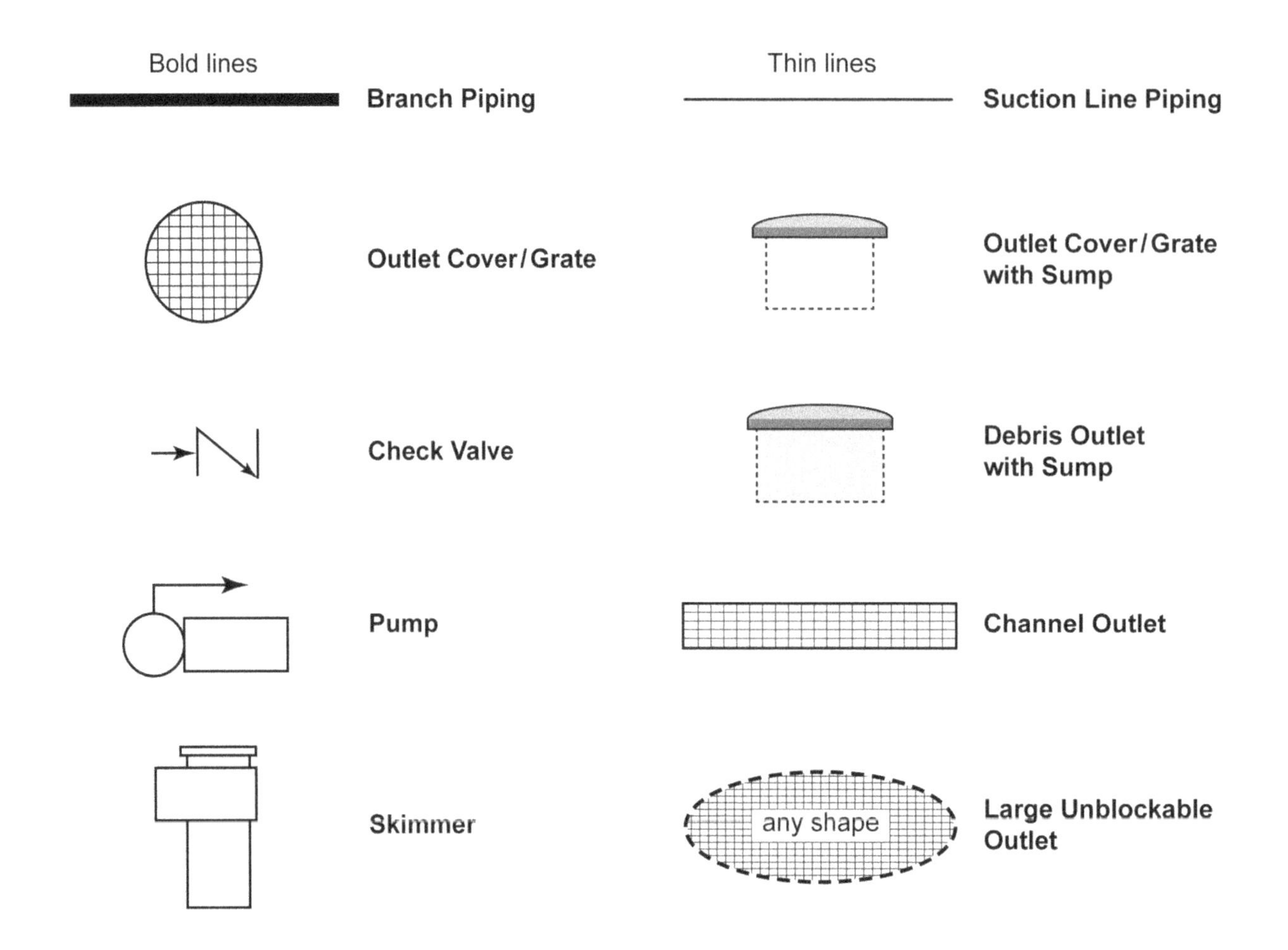

Appendix B: Field Checklist for Identifying Suction Entrapment Hazards

This Appendix is not part of the American National Standard ANSI/APSP/ICC-7 2013 but is included for information only. Additional copies of the ANSI/APSP/ICC-7 standard and this Appendix can be purchased by contacting APSP Member Services at 703.838.0083, ext. 301.

Introduction

This field checklist for identifying suction entrapment hazards provides information and a systematic process that will help identify and eliminate suction entrapment hazards in swimming pools, wading pools, spas, hot tubs, and catch basins. This information and system is intended to address the hazards of hair entrapment, limb entrapment, body suction entrapment, evisceration/disembowelment, and mechanical entrapment. It does not replace or supersede the information in the body of the ANSI/APSP/ICC-7 standard. These guidelines are intended for use in inspecting, maintaining, and upgrading residential and public swimming pools, wading pools, spas, hot tubs, and catch basins. They are appropriate for use by service companies, builders, installers, facility owners/operators, home inspection specialists, parks and recreation personnel, and others who are responsible for pool and spa safety.

Reference numbers next to each block are used to facilitate telephone discussion. Mark the tracking boxes with an × to clearly document the current condition and actions needed and/or taken.

> ⚠ **DANGER: To avoid serious injury or death, close the pool or spa to bathers if any suction outlet cover/grate is missing, broken or inoperative.**

Company ______________________

www. ______________________

Address ______________________

City ______________________

State ____________ Zip ____________

Date ____________ Phone ____________

Inspected by ______________________

Pool ______________________

Pump System ______________________

Address ______________________

City ______________________

State ____________ Zip ____________

Date ____________ Phone ____________

Owner/Operator ______________________

EVALUATION / ACTIONS TAKEN

____

____

____

____

____

____

____

____

____

____

____

Inspector ______________________
(Print Name)

____
(Signature) (Date)

Owner/Operator ______________________
(Print Name)

____
(Signature) (Date)

The provisions described herein are not intended to prevent the use of any alternative configuration or system, provided any such alternative meets the intent and requirements of these Guidelines.

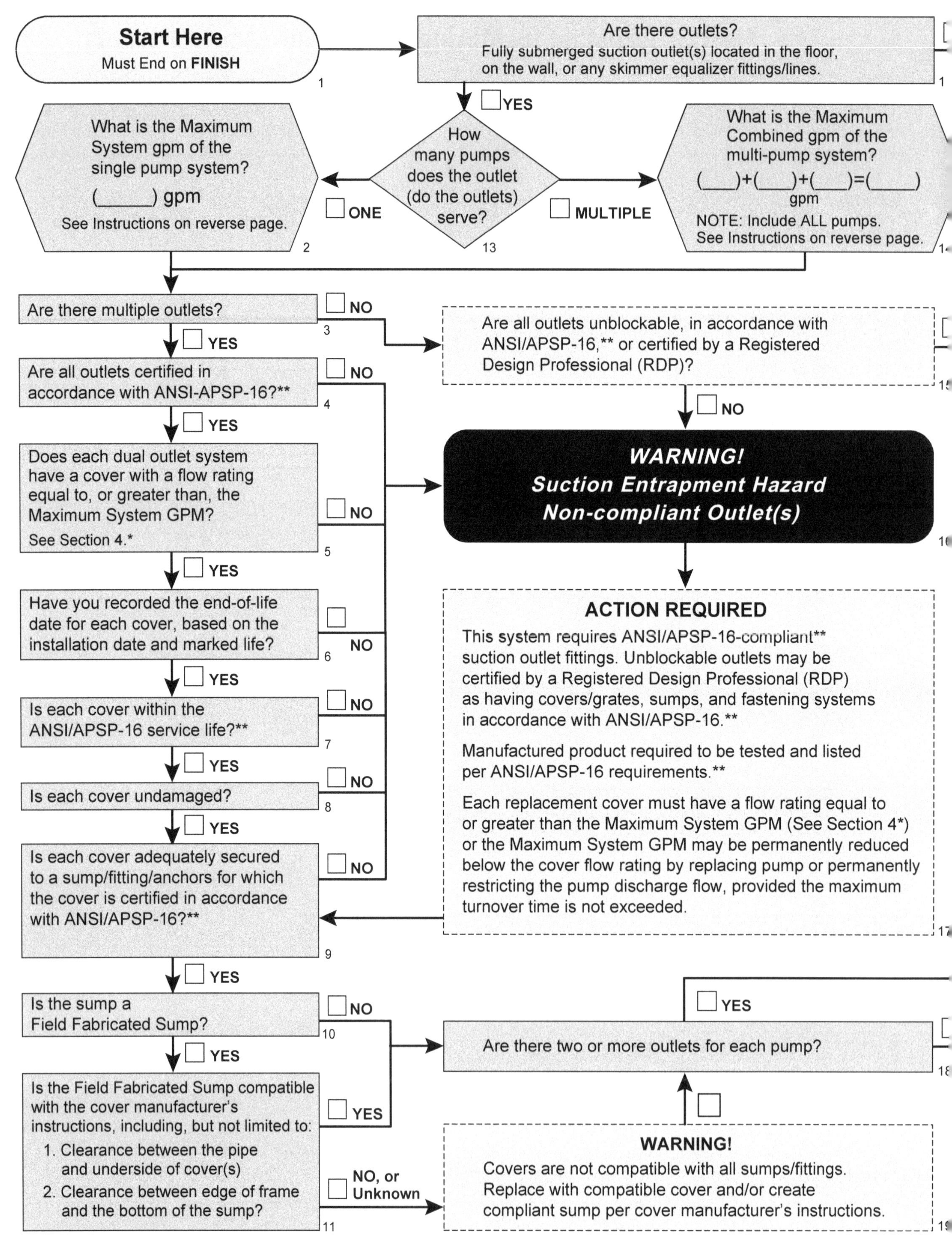

* Unless explicitly noted, all section numbers refer to ANSI/APSP/ICC-7 2013.

** All references to ANSI/APSP-16 mean ANSI/APSP-16

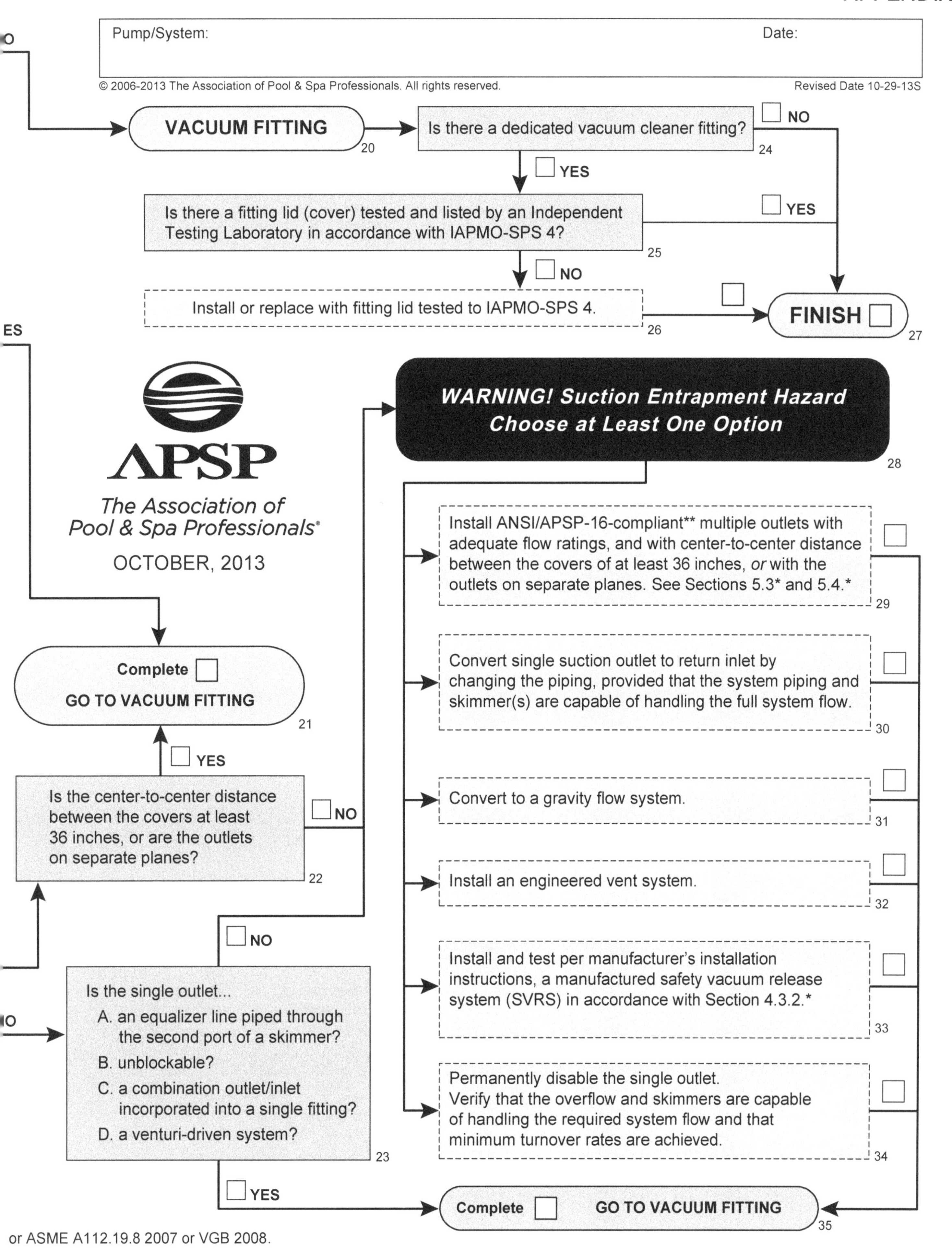

or ASME A112.19.8 2007 or VGB 2008.

Finding the Maximum Flow Rate of an Existing System

Preparation:

1. Open all valves to their full open position for pool or spa circulation. (For secured systems, do not adjust valves.)
2. Remove eyeball fittings from return inlets (when removable by hand).
3. Clean skimmer and pump baskets. Turn off skimmer to isolate outlet, if possible.
4. Backwash or clean sand filter/DE grids, or remove cartridge.

When inspecting existing installations, the maximum possible flow rate of suction system must be determined as explained in 4.4.9.*.

Pump Method 1: Measure flow rate with a flow meter accurate to ±10% (see Section 4.4.9).*

Pump Method 2: Calculate using pressure and vacuum gauge readings (see diagram below).

1. Install a vacuum gauge as close to the bottom of the strainer basket as possible.
2. Install a pressure gauge as close to the pump discharge as possible.
   NOTE: It may be necessary to use an NPT[9] × barb fitting with a short section of plastic tubing connected to a gauge if gauges cannot be screwed into drain holes provided in pump.
3. Multiply vacuum reading by 1.13 and record.
4. Multiply pressure reading by 2.31 and record.
5. Add results of steps 3 and 4 together to get the approximate Total Dynamic Head (TDH) in feet of head.
6. Using the published curve for the pump, find the Total Dynamic Head calculated above on the vertical axis, and read the flow rate on the horizontal axis.
7. This will give you the maximum flow rate within approx. 10%.

Pressure head: gauge psi × 2.31 = feet of water
Suction head: gauge inches Hg × 1.13 = feet of water

EXAMPLE: If the pressure gauge reads 14 psi and the vacuum gauge reads 6 inches of mercury (Hg), the approximate Total Dynamic Head (TDH) of the system would be 39.12 feet.

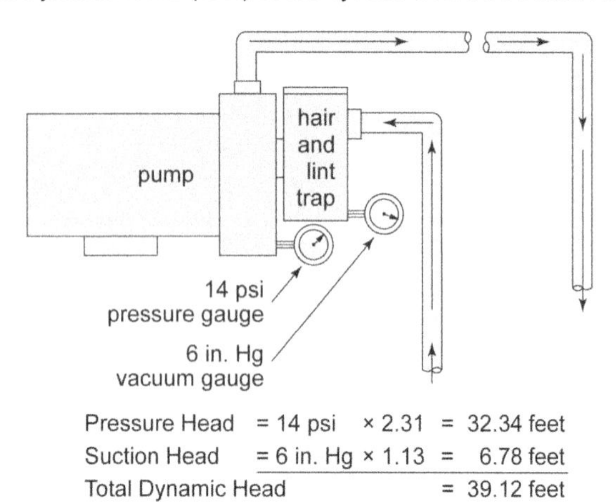

Pressure Head	= 14 psi × 2.31	= 32.34 feet
Suction Head	= 6 in. Hg × 1.13	= 6.78 feet
Total Dynamic Head		= 39.12 feet

Gravity Flow Calculation

$$\text{Flow (gpm)} = \sqrt{\frac{1786 \times [D\ (\text{inch})]^5 \times H\ (\text{inch})}{L\ (\text{inch}) + [55 \times D\ (\text{inch})]}}$$

(Where 55 D accounts for energy loss of stream)

EXAMPLE: Gravity flow through 2" IPS Schedule 40 PVC pipe with an inside diameter of 2.067" with 32.0 feet of pipe and 2 elbows of equivalent length of 6.0 feet. The top of the pipe opening into the collector tank is 8" below pool water level.

$$\text{Flow (gpm)} = \sqrt{\frac{1786 \times [2.067]^5 \times 8}{[32 + (2 \times 6)] \times 12 + [55 \times 2.067]}} = 29 \text{ gpm}$$

Cover/Grate Audit

Existing Pump ______________ (Manufacturer) ______________ (Model)

Pool Volume ______________ (Gallons)

Filter ______________ (Manufacturer) ______________ (Model) ______________ (Size (Sq. Ft.))

Existing Cover ______________ (Manufacturer) ______________ (Model) ______________ (GPM)

Pressure ______________ (PSI) Vacuum ______________ (Inches of Hg)

TDH ______________ (Feet of water) System Flow ______________ (GPM) (from Pump Curve)

Maximum Flow ______________ (GPM)

New Cover ______________ (Manufacturer) ______________ (Model) ______________ (GPM)

Replacement Date _____/_____/ _____

Maximum Drawdown ______________ (Calculated)

______________ (Measured) ______________ (Measured) ______________ (Measured) ______________ (Measured)

NOTE: Check cover manufacturer's installation instructions for the following items per ANSI/APSP-16.**

- ☐ Cover compatible with sump
- ☐ Attachments (hardware/screws)
- ☐ Field fabricated sump as specified by cover manufacturer

The Association of Pool & Spa Professionals®

8. National Pipe Thread

Appendix C: Entrapment Avoidance Warning Sign

This appendix is not part of the American National Standard ANSI/APSP/ICC-7 2013. It is included for information only.

Drowning Hazard

Avoid Drain Covers

Avoid Body Entrapment

Avoid Evisceration

Avoid Hair Entanglement

Avoid Finger Entrapment

- **Never play or swim near drains (submerged suction fittings). Your body or hair may be trapped, causing permanent injury or drowning.**
- **Never enter the pool or spa if a drain cover (suction fitting cover) is loose, broken, or missing.**
- **Immediately notify the pool/spa owner or operator if you find a drain cover (suction fitting cover) loose, broken, or missing.**

For further information contact The Association of Pool & Spa Professionals.

Visit the U.S. Consumer Product Safety Commission website at www.cpsc.gov to read their entrapment guidelines: "Guidelines for Entrapment Hazards: Making Pools and Spas Safer 2005"
Also visit: www.poolsafely.gov and www.APSP.org

> IMPORTANT SAFETY NOTE: If you choose to display this warning device as a sign, please make sure that it conforms to ANSI/NEMA Z535.4-2011 Standard for Product Safety Signs and Labels, or latest revision.

Appendix D: Sources of Material

This Appendix is not part of the American National Standard ANSI/APSP/ICC-8 2005 (R2013). It is included for information only.

ANSI American National Standards Institute
25 West 43rd Street
New York NY 10036
Tel: 212-642-4900
Fax: 212-398-0023
www.ansi.org

APSP Association of Pool & Spa Professionals
(formerly National Spa and Pool Institute)
2111 Eisenhower Avenue
Alexandria VA 22314
Tel: 703-838-0083
Fax: 703-549-0493
www.APSP.org

ASTM International Standards Worldwide
(formerly American Society of Testing & Materials)
100 Barr Harbor Drive
West Conshohocken, PA 19428-2959
Tel: 610-832-9500
Fax: 610-832-9555
www.astm.org

IAPMO International Association of Plumbing and Mechanical Officials
5001 E. Philadelphia Street
Ontario, CA 91761
Tel: 909-595-8449
Fax: 909-472-4150
www.iapmo.org

NFPA National Fire Protection Association
1 Batterymarch Park
Quincy MA 02269
Tel: 617-770-3000
Fax: 617-770-0700
www.nfpa.org

UL Underwriters Laboratories Inc.
333 Pfingsten Road
Northbrook IL 60062-2096
Tel: 847-272-8800
Fax: 877-272-8129
www.ul.com

(Approved by the American National Standards Institute October 8, 2013)

ANSI/APSP/ICC-7 2013

Approved American National Standard ANSI

American National Standard for Suction Entrapment Avoidance in Swimming Pools, Wading Pools, Spas, Hot Tubs, and Catch Basins

Familiarity with the ANSI/APSP/ICC standards is essential for anyone who builds, manufactures, sells, or services pools, spas or hot tubs.

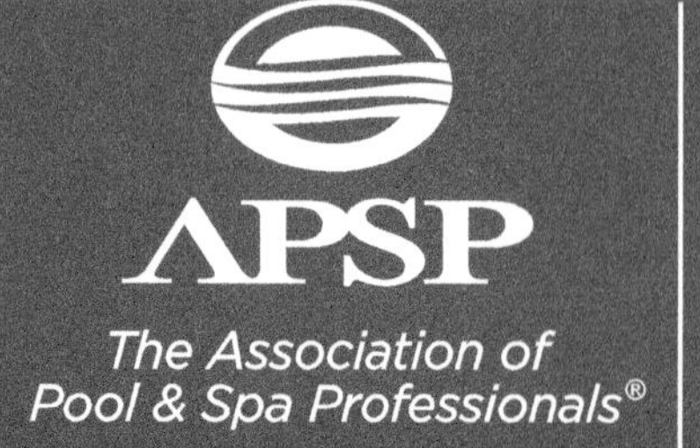

2111 Eisenhower Avenue
Alexandria VA 22314-4695

703.838.0083
memberservices@APSP.org
APSP.org